SOVIET RESEARCHES ON LUMINESCENCE

ISSLEDOVANIYA PO LYUMINESTSENTSII

ИССЛЕДОВАНИЯ ПО ЛЮМИНЕСЦЕНЦИИ

Transactions (Trudy) of the P.N. Lebedev Physics Institute, Volume XXIII

SOVIET RESEARCHES
ON LUMINESCENCE

Edited by

Acad. D. V. Skobel'tsyn

Authorized translation from the Russian

CONSULTANTS BUREAU
NEW YORK
1964

The original text was published by the USSR Academy of Sciences
Press in Moscow in 1963 as Volume XXIII of the Transactions
(Trudy) of the P. N. Lebedev Physics Institute.

Исследования по люминесценции
Труды Физического института им. П. Н. Лебедева, Том XXIII

Library of Congress Catalog Card Number 64-7763

ISBN 978-1-4615-8548-0   ISBN 978-1-4615-8546-6 (eBook)
DOI 10.1007/978-1-4615-8546-6

CONTENTS

THE ELECTROLUMINESCENCE OF CRYSTALS

A. N. Georgobiani

INTRODUCTION

It is well known that luminescence is the term used to describe the excess radiation from a body over and above the thermal radiation and persisting for a time which greatly exceeds the period of a light vibration. The first half of this definition, proposed by Wiedemann, distinguishes luminescence from equilibrium thermal radiation; the second half, introduced by Vavilov, distinguishes luminescence from various forms of scattering and from induced radiation, such as Vavilov-Cherenkov radiation, etc.

Distinctions are made between photo-, cathodo-, x-ray-, and other forms of luminescence, depending on how energy is introduced into the luminescent body.

Electroluminescence is the name given to that form of fluorescence in which the radiating body receives energy directly from an electric field. It should be noted that luminescence under the influence of cathode rays is not called electroluminescence, because in this case the necessary energy is not supplied directly from the electric field to the radiating body but by means of extraneous electrons.

Electroluminescence of gaseous bodies (radiation from a gas discharge) has been known for a long time and is widely used in luminescent lamps and gas discharge tubes.

In 1923 Losev [1] observed radiation from silicon carbide crystals when a voltage was applied to them directly.

In 1936 Destriau [2] detected radiation from finely crystalline zinc sulfide, activated with copper, when this material was suspended in a liquid dielectric and placed between the plates of a condenser, to which an alternating electric voltage was applied. For a long time it was not certain whether or not this radiation was photoluminescence excited by the ultraviolet light of a corona discharge, itself excited by the applied voltage. Only in the nineteen fifties [3, 4], as the result of spectral investigations and the direct comparison of radiations from the same luminophor in the presence and absence of a corona discharge, was it established that electroluminescence is a separate process on its own. Reasonably bright electroluminophors were synthesized during the same period, and intensive investigation of this phenomenon began.

An electric field can excite crystals directly, owing to the tunneling transfer of electrons from the valence band and from luminescence centers to the conductivity band (the Zener effect), or indirectly, by acceleration of electrons in the electric field to energies sufficient to ionize the crystal lattice and luminescence centers (impact ionization). Impact excitation of luminescence centers can also occur. Similar processes evidently take place in most electroluminophors and in p-n junctions connected in the blocking-off direction.

An electric field can cause crystals to fluoresce in another way. The point is that crystal phosphors are mostly semiconductors with impurity carriers. If these are in direct contact with electrodes, then it is possible that additional minority charge carriers may be introduced. The radiation, arising from the recombination of these with majority carriers, is a form of luminescence usually called injection electroluminescence. Broadly speaking, this form of luminescence should be classified as electrothermal luminescence. In fact, the free electrons and holes giving rise to such luminescence are produced by thermal vibrations of the lattice. The action of the electric field in this case is to bring the free electrons and holes together, so that they have a chance to recombine. Such luminescence is observed with p-n junctions connected in the positive direction.

It is normal practice in the literature to classify the electroluminescence of crystal phosphors under the Losev or Destriau effects, according to whether the crystals are or are not in direct contact with the electrodes. In the first case there is the possibility that mobile charge carriers can penetrate from the electrodes to the luminophor, but this is impossible in the second case because there is a layer of dielectric between the phosphor crystals and the electrodes.

The main directions for the practical application of the electroluminescence of solid bodies are as follows:

1. The construction of illuminating devices without gas-filled or vacuum tubes. Electroluminescent lamps have already been made which show only a 20-40% decrease in brightness after ten thousand hours [5].

2. The production of various types of light amplifier and light converter. Light amplifiers with amplification coefficients of up to 100 are now being produced [5].

3. The construction of instruments which transform invisible radiation or convert it into visible radiation.

4. The construction of television screens in which excitation is produced by an electric field, applied to the screen and modulated in an appropriate way. Such a screen has various advantages over the type now used for television, in which radiation is excited by a modulated electron beam, accelerated in a vacuum tube to an energy of about 10 keV. An electroluminescent screen does not require such a vacuum tube, and the excitation voltage required is less by a factor of 1/10 to 1/100.

The most widely used luminophors are synthesized on a zinc sulfide basis. The best have a luminance of up to 1000 apostilbs, and their luminous efficiency is about the same as that of an incandescent lamp [6].

CHAPTER I

GENERAL PROBLEMS
OF ELECTROLUMINESCENT CRYSTALS

§ 1. Ionization by the Electric Field

The mechanism of electron transfer to the conductivity band by tunneling through the potential barrier was first considered by Zener [7], and the effect carries his name. He obtained an equation for the probability of transfer of a valence electron to the conductivity band in unit time:

$$P_t = \frac{eEd}{2\pi\hbar} \exp\left\{ - \frac{\pi}{4ehE} \sqrt{2m^*}\, \Delta^{3/2} \right\},$$

(I.1)

where E is the value of the electric field; e and m* are the charge and effective mass of an electron; Δ is the width of the forbidden band; d is the constant of the crystal lattice. However, this equation is only valid when

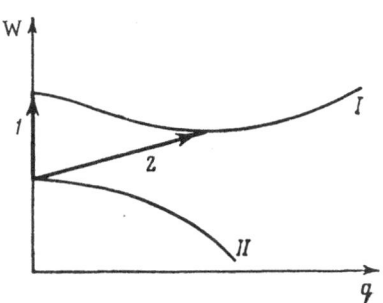

Fig. 1. The relation between the electron energy W and its quasi-momentum q; curve I is for the conductivity band and curve II for the valence band. The arrows 1 and 2 denote direct and indirect transitions respectively.

the quasi-momentum q of an electron is the same at the bottom of the conductivity band and at the top of the valence band. This condition is often not fulfilled so that, because of the law of conservation of momentum, transitions are only possible when, although the energy level is higher at the bottom of the conductivity band, the quasi-momentum of an electron is the same there as at the top of the valence band. In this case, instead of the forbidden band width Δ in equation (I.1), it is necessary to insert the difference between this level and the top of the valence band. However, if we are dealing with a body such that it is possible to transfer the requisite momentum to an electron, then there is also a possibility of electron transfer to the bottom of the conductivity band. This body can be the crystalline lattice itself (absorption or creation of phonons). Such a transition is called indirect [8] (Fig. 1). In such cases indirect transitions occur with a high probability, while direct transitions are not favored.

Indirect transitions have been investigated theoretically [9]. The probability of an indirect transition in unit time increases with temperature owing to the increase in the number of phonons. In the temperature region

$$T_0 < T < T_D.$$

(I.2)

where T_D is the Debye temperature and $T_0 \simeq 10^{-4}$ E (E is the field in volts per centimeter), the probability is given by the equation

$$P_t \approx a_0 \frac{1}{e^{\frac{\hbar\omega_q}{kT}} - 1} \exp\left\{ - \frac{4\sqrt{2m^*}\,(\Delta \pm \hbar\omega_q)^{3/2}}{3\hbar eE} \right\},$$

(I.3)

where a_0 is a factor that depends slightly on the field and temperature; $\hbar\omega_q$ is the energy of a phonon required to fulfill the law of conservation of momentum; k is the Boltzmann constant; the signs + or − depend on whether the transition occurs with creation or absorption of a phonon.

The probability of ionization by the field increases with temperature still further because of the reduction in width of the forbidden band as the crystal is heated. The relative change in Δ with temperature is of the same order as the coefficient of linear expansion of the crystal lattice. For germanium $d\Delta/dT \simeq -4 \times 10^{-4}$ eV per degree [10].

If an electron absorbs a phonon for a tunneling transition, then in addition to momentum it also acquires energy, which is equivalent to a reduction in the height of the potential barrier by the same amount [see equation (I.3)]. The greater the absorption of phonons, the more the "reduction" of the barrier. Thus although the absorption of several phonons is very much less probable than the absorption of one, tunneling transitions with the aid of several phonons may be effective at high temperatures and with weak fields. The tunneling transition with participation of several phonons has been investigated theoretically [9].

Equation (I.3) involves the product of the probability of acquisition of energy $W = \hbar\omega_q$ from the crystal lattice and the probability of penetration through a potential barrier reduced by this amount of energy. In order to calculate the full probability of tunneling ionization of the crystal lattice, it is necessary to sum (I.3) for all the energies which an electron may obtain from the lattice. In this case the increase in probability of penetration for the absorption of each additional phonon depends on the field, so that the number of phonons absorbed, which represents the largest term in the penetration probability, also depends on the field. Therefore, the overall equation for tunneling ionization with the participation of several phonons appears quite different from the usual equation for ionization by a field (I.1; I.3). According to [9], at values of the field for which

$$E < \frac{kT}{ed} ,$$

(I.4)

we can use the equation

$$P_r^T = a \exp\left\{ -\frac{\Delta}{kT} + \frac{1}{24}\frac{(e\hbar E)^2}{m^*(kT)^3} \right\} ,$$

(I.5)

where a, like a_0, varies slightly with field and temperature.

It is useful to note that equation (I.5) is in a sense intermediate between the equation for probability of thermal ionization and equations (I.1; I.3) for the probability of direct ionization by the field (the Zener effect).

If we consider the ionization of a local center whose level is separated from the band by Δ_c, then in all these equations it is necessary to use the effective dimension of this center instead of the crystalline lattice constant. The effective dimension d_c of a local center whose energy level is located substantially above the middle of the forbidden band is approximately given by

$$d_c \approx \frac{\hbar}{\sqrt{2m\Delta_c}} .$$

(I.6)

§ 2. Impact Ionization

A free charge is accelerated under the influence of a powerful electric field and may excite or ionize atoms by collision. A similar phenomenon is observed in a crystal with mobile "free" charge carriers in the presence of a strong electric field. The term "free" in this case is provisional in character, because such charges move about in a nonhomogeneous field. However, if the field heterogeneity is strictly periodic, then the movement is similar to that of a free charge. The detailed process of impact ionization in a solid body takes place as follows. When a strong electric field is applied, the conduction electrons from the conductivity band acquire energy which is sufficient for them to eject electrons from the valence band; these ejected electrons then contribute to the conductivity and are transferred to the conductivity band. Impact ionization of impurity centers also occurs in a solid body, and this does not differ essentially from impact ionization of an isolated atom. At some critical value of the voltage of the electric field, this process leads to such a steep increase in current density that electric breakdown of the semiconductor occurs.

Under real conditions a crystalline lattice has defects. In addition, at temperatures above the absolute zero it is subject to thermal vibrations, which also disturb its spacing. The charge carriers, in their motion through the crystal, are scattered at such disturbed spacings of the crystalline lattice. When this happens, an electron, gathering energy from the field, gives it up to the lattice. Thus scattering interferes with the accumulation by an electron of sufficient energy for impact ionization. A specially pronounced effect on the movement of charge carriers is shown by vibrations of the crystalline lattice in which the accompanying ions or atoms are displaced from their equilibrium positions in the antiphase (optical phonons), since these displacements give rise to the greatest heterogeneity. The value of this heterogeneity, which determines the extent of the disturbing action of such vibrations on the movement of charge carriers, is itself determined by the type of bond in the crystalline lattice.

In the substances which concern us there are two types of bond, ionic and covalent. In the case of ionic bonds, positive and negative ions are located in a definite order at sites in the crystalline lattice, and the main part of the bond energy is determined by the Coulomb interaction of the ions. In the case of covalent bonds, there is a collectivization of the valence electrons of atoms joined into molecules, and the main part of the bond energy is the exchange energy of joint electron pairs.

In any specific crystal there is an intermediate type of bond, which can be characterized approximately by the percentage of ionic and covalent bonds.

The intensity of the electric field arising from the thermal vibrations of the crystal lattice of an ionic crystal is very large; it is usually larger than the field intensity produced by application of an external voltage to the crystal. Because of this, the mean free path of an electron in an ionic crystal, determined by scattering due to the thermal vibrations of the lattice, is of the same order as the crystalline lattice constant.

At low speeds the electron deforms the lattice as the result of polarization, and the deformation region moves through the crystal with the electron. The electron is then said to be in a polaron state [11]. The effective mass of the polaron is normally several hundred times that of an electron, so that it is considerably more difficult to accelerate it by means of an external electric field.

Owing to strong interaction with the lattice in ionic crystals, the electrons are in thermal equilibrium with the latter even in very high fields. According to [12], the mean energy of an electron in fields up to breakdown is only a few times kT. Very few electrons are able to accumulate sufficient energy for impact ionization. For the most part, the electrons expend the energy acquired from the field by warming up the crystal. The result is that thermal disruption of such crystals can occur even with fields which are insufficient to give appreciable electroluminescence.

In crystals with covalent bonds, the fields produced by thermal vibrations are very much less than in the case of ionic crystals. Accordingly, the interaction of electrons with the thermal vibrations of such a lattice is normally weaker than in the case of an ionic crystal, so that the mean free path of an electron in a crystal with covalent bonds is greater, by several orders of magnitude, than in an ionic lattice. The electrons in such crystals are not in thermal equilibrium with the lattice when a high voltage is applied, i.e., many of the electrons have an energy considerably exceeding the mean thermal energy kT ("hot" electrons*). For this reason the process of impact ionization occurs at considerably lower fields in crystals with covalent bonds. According to [18], the mean energy of the electrons in a covalent crystal is related to the electric field by the equation

$$W = 0.64 \ W_{imp}\left(\frac{E}{E_{imp}}\right)^{\frac{3}{2}}, \tag{I.7}$$

---

* Recombination of "hot" charge carriers evidently also explains the emission of light quanta with energies greater than the width of the forbidden band in the recombination fluorescence from covalent or almost covalent semiconductors: Ge [13], Si [14, 15], SiC [16], InSb [17].

where W_{imp} is the value of the threshold energy for ionization, * and E_{imp} is some characteristic value of the field:

$$E_{imp} = \left\{ \frac{3\hbar\omega_0 W_{imp}}{4e^2 l_0^2} \frac{e^{\hbar\omega_0/kT} + 1}{e^{\hbar\omega_0/kT} - 1} \right\}^{1/2} , \tag{I.8}$$

where l_0 is the mean free path of a thermal energy electron; $\hbar\omega_0$ is the energy of a longitudinal optical phonon.

The mean probability of impact ionization by an electron in unit time, for the atoms or ions of a crystalline lattice with impurity centers present in a field E, calculated for one particle is given by [18, 19]

$$P_{imp} = \sigma_0 \left(\frac{E}{E_{imp}} \right)^n e^{-\left(\frac{E_{imp}}{E} \right)^2} , \tag{I.9}$$

where σ_0 is the order of magnitude of the effective cross section for interaction with impact ionization, i.e., about 10^{-16} cm^2, and various values have been given for n lying between 1 and 5.

From equation (I.9) we can easily obtain a coefficient of impact ionization, i.e., the number of ionizations produced by an electron moving in a field E along a path of unit length. To do this it is necessary to divide P_{imp} by the mean speed of the electron.

Seitz [20] obtained an equation for the probability of impact ionization in an ionic crystal:

$$P_{imp} = \sigma_0' e^{-\frac{E_{imp}'}{E}} , \tag{I.10}$$

where E_{imp}' is also some characteristic value of the field. In his calculations he started from a simple model and supposed, indeed, that the electron, owing to its fluctuations, could traverse in the direction of the field a path l_E long enough for it to acquire ionization energy without collision with the lattice. The probability of such a fluctuation is $(-l_E/\bar{l})$, where $\bar{l}$ is the mean free path. However, Seitz did not allow for the fact that the mean free path of an electron in an ionic crystal increases as the energy of the electron increases, because an increase in energy reduces the effectiveness of interaction between the electron and the lattice vibrations. According to [21] the mean free path is proportional to the electronic energy. For this reason equation (I.10) has been found imprecise.

A more correct equation for the mean probability of impact ionization can be obtained by solving the kinetic equation. The treatment of this problem by solution of the kinetic equation in [22] showed that the Seitz equation was only applicable to weak fields such that

$$E \ll E_1^* = \frac{2kT}{\epsilon l_0} . \tag{I.11}$$

With such fields, only those electrons are effectively accelerated which have obtained, from the thermal energy of the crystal, sufficient energy to overcome the retarding action of the lattice. For stronger fields, which themselves determine the number of "hot" electrons, a relation of the type (I.9) applies.

Under normal conditions impact ionization occurs at lower field strengths than does field ionization. For this reason cascade breakdown of a crystal can take place in fields which are still insufficient for strong field ionization. However, the development of breakdown is retarded in thin films, since it is then necessary that a

_____

* W_{imp} is normally greater than the width of the forbidden band Δ. For equal effective masses of electron and hole, and for spherically symmetrical energy zones, it follows from the laws of conservation of energy and momentum that $W_{imp} = (3/2)\Delta$.

large field should extend over a large enough region, and field ionization may effectively compete with the process of impact ionization.

§ 3. Concentration of the Electric Field in Electroluminescence

The process of ionization of the crystalline lattice and local centers by the electric field, both directly and by impact ionization, depends very much on the field strength. It is, therefore, natural that any region of the crystal where the field strength is, for any reason, higher than in the other parts should contribute most to the electroluminescence.

As the result of concentration of the electric field in electroluminescent crystals, most of these show electroluminescence at voltages corresponding to the mean field in the crystals, which are still well below the breakdown values.

The electric field is high in those regions of the crystal where the specific resistance is greatest. The other parts of the crystal play the part of a ballast resistance, preventing breakdown. In fact, since an approach to the breakdown field is only achieved in those regions where the internal voltage is concentrated, the conductivity of these regions increases, and the field ceases to concentrate there. Thus the field either becomes uniformly distributed over the crystal or concentrates in those regions which now have the highest specific resistance. This movement of the strongest part of the field, accompanied by a movement of the region giving the strongest radiation as compared with the mean value for the whole crystal, has been observed with CdS monocrystals [23, 24].

On the other hand, in order to excite electroluminescence from a crystal with a uniform field distribution, it is necessary to establish a prebreakdown state over its whole volume. This can be achieved only with a very uniform sample, since otherwise the breakdown state will first be reached in the part of the crystal with the lowest electric strength, and this will then be removed from the system.

It is clear from the above that the condition of field concentration has great importance for the practical use of electroluminescent materials. It should be added that most electroluminescent materials radiate very heterogeneously with respect to the volume of the crystals; some parts are brighter by several orders of magnitude than the main mass of the crystal [14, 16, 23, 25, 26].

In different cases there may be various causes which can lead to concentration of the electric field. Some of these are discussed below.

The External Shape of the Sample. It is well known that changes in the shape, dimensions, and relative positions of the condenser electrodes can produce pronounced heterogeneity of the electric field. Similar results can be produced by irregularities in the form of the crystals placed in the condenser (sharp point, etc.). An estimate has been made [27] of the possible field heterogeneity from this cause and its effect on the electroluminescence of ZnS crystals.

It has been stated that a region where the points of the crystals are oriented along the field luminesce most strongly [25].

The Depletion Barrier of the Mott-Schottky Type and the p-n Junction. If a crystal, isolated from the electrodes, is placed in an electric field, then the free charge carriers, depending on their sign, move away from their respective edge of the crystal and form a space charge of opposite sign. This charge screens the interior of the crystal from the external electric field, which is then practically restricted to the surface region from which the free charge carriers have been removed. It is in this region, with a reduced concentration of free charge carriers, known as a "depletion" barrier of the Mott-Schottky type, that there occurs the main gradient of the voltage applied to the crystal.

The process of field concentration takes place quite rapidly when the levels of the donor or acceptor, which determine the conductivity of the crystal, are low enough so that they can be rapidly ionized thermally. With germanium and silicon rectifiers at room temperature, the depth Δ_D is less than kT, so that they are practically all ionized at room temperature.

With electroluminophors of the ZnS type, the level is very much deeper than kT, but, as will be seen in the next chapter, the same process still occurs because, at the field strengths attained with these crystals during electroluminescence, there is intensive liberation of charge carriers under the influence of the electric field. This is either by the tunneling effect or by impact ionization. In other words, with these crystals the process of concentrating the electric field can be achieved by the field itself. In the Mott-Schottky model it is proposed that the concentration of ionized donors and acceptors does not alter over the width of the "depletion" barrier. The relation between the width L and the applied voltage U can be found by solving the Poisson equation, and is given by the expression

$$L = \left[\frac{\varepsilon U}{2\pi e^2 N_C}\right]^{1/2},$$

(I.12)

where N_C is the concentration of charge centers creating the space charge in this barrier; ε is the dielectric constant of the crystal; U is the voltage; e is the charge on an electron. The field in the barrier decreases linearly with the distance from the boundary:

$$E(x) = -\frac{2U}{L}\left(1 - \frac{x}{L}\right).$$

(I.13)

Here x is the distance from the boundary. It is clear from equations (I.12) and (I.13) that the maximum value of the field in the barrier is proportional to $\sqrt{U}$.

A similar phenomenon occurs with a p-n junction connected in the blocking direction. In this case the mobile charge carriers move away from the boundaries of the p-n junction and thus create a region round it depleted of carriers, similar to that described above. The electric field is concentrated in this region. Equations similar to those above are obtained if the transition from the region with hole conductivity to the electron region is sufficiently sharp.

It follows from the above that the greatest amount of radiation should come from the region where the field is concentrated. In practice, with semiconductor rectifiers connected in the blocking direction, the bulk of the radiation comes from the region of the p-n junction [13-16].

It should be noted that some authors (see, e.g., [27]) have expressed the view that field concentration at p-n junctions and "depletion" barriers are impossible in the case of electroluminescence from isolated crystals. These authors maintain that the field at such a barrier cannot be greater than the field in the dielectric film between the crystals, which they say is approximately equal to the mean field in the electroluminescent condenser. However, allowing for the fact that this dielectric film may be very much thinner than the total thickness of the electroluminescent layer, it is possible for the field in the barrier to be many times greater than the field in this film.

Other authors [28, 29], in calculating the proportion of the voltage acting across the crystals, also started on the basis of a mean field in the electroluminescent condenser and came to the conclusion that the field is indeed concentrated in the crystals but in a very small part of their volume, and that there acts across this volume only about a tenth of the total voltage applied to the crystals. However, for the excitation of radiation in a crystal, it is not only the field strength that is important but also the applied voltage. The above authors clearly ignored this point. In fact, for the excitation of electroluminescence in ZnS crystals it is essential that the voltage acting on the crystals should be at least about 3 V, since light quanta with an energy of about 3 eV are emitted during the electroluminescence. Fluorescence from an electroluminescent condenser has been described [30] with an applied voltage of only approximately 1.5 V.* If the authors of [28, 29] are correct, then the voltage in the region of field concentration will have dropped to 0.15 V, which is clearly insufficient for ionization or excitation of radiation centers.

---

* Electroluminescence at such low voltages will be considered in detail below.

An analysis which appears in the next chapter shows clearly that it is those crystallites between which there is a thin enough dielectric film which are excited; the major part of the total voltage applied to the condenser acts across these crystallites. It is therefore more correct in calculations to start, not from the mean field in the electroluminescent condenser, but from some voltage acting across the electroluminescing crystals.

The Boundary Region of Ferroelectric Materials. Because of ferroelectric polarization, the dielectric constant of a ferroelectric material can reach a very high value, depending on the temperature of the sample; for example, it will be of the order of 1000 at the Curie temperature. However, at the sample surface the conditions for ferroelectric polarization are disturbed, and there the dielectric constant is optical in character, i.e., of the order of a few units. When such a sample is exposed to an alternating voltage of such a high frequency that the impedance of the sample is practically independent of its conductivity, a strong field is set up in the boundary region. This field exceeds that in the interior of the sample by a factor which is equal to the inverse ratio of the dielectric constants. Electroluminescence has been observed [26] from the boundary regions of samples of $BaTiO_3$, $SrTiO_3$, $KNbO_3$, $CaTiO_3$, and TiO_2. When the temperature was altered, the luminance varied in proportion to the change in ratio of the dielectric constants at the center and at the edge.

Other Phenomena Causing Field Concentration. It has been concluded [31] that conditions for field concentration (though less intense than at p-n junctions) can occur at the boundaries separating two crystal phases which differ in conductivity, in extent of alloying (n-n', p-p' junctions), or in crystal structure. An example is the boundary separating the cubical and hexagonal modifications in ZnS crystals.

§ 4. Electroluminescent Materials

The emission of light quanta is the last stage in any type of luminescence, so it would be anticipated that there would be a relation between the capacities of crystals for photo-, cathodo-, x-ray-, radio-, and electroluminescence. However, it does not follow that this relation should be well defined, since for each type of luminescence it is a different form of energy which is being supplied to the crystal. The crystalline body must accept the energy and, under its influence, be converted to an excited state; it is the return from the excited state which may be accompanied by the emission of light quanta.

In order that crystal phosphors should act as good electroluminophors, it is not sufficient that they should contain radiation centers. It is also necessary that there should be a movement of charges in the crystal when it is placed in an electric field, since otherwise the field cannot give up energy to the crystal. Moreover, the movement of charges must lead to excitation of the crystal, either directly by the field or by impact ionization. In this process, as noted in the preceding section, it is desirable that regions should exist in the crystal where the electric field is considerably more intense than its mean value. The type of crystal bond has a pronounced effect on the possibility of the occurrence of such conditions. For example, it is more difficult to excite electroluminescence in ionic crystals than in crystals with covalent bonds. The reasons for this are that in ionic crystals:

1. There is a short mean free path for electrons and holes, due to their scattering by phonons.

2. There is a relatively wide forbidden band.

3. There is a low concentration of free charge carriers so that, as a rule, regions where the field is concentrated (p-n junctions, "depletion" barriers, etc.) are not formed.

For these reasons, ionic crystals require higher fields than covalent crystals for the excitation of electroluminescence. The disintegration of such crystals by local breakdown, as for instance at the surface, can occur with fields which are still inadequate for significant electroluminescence. Also, as has already been noted, the mean free paths of electrons and holes in ionic crystals are so small that most of the energy acquired by them from the field is degraded in heating the crystals. For this reason such crystals may also be disintegrated by heat at fields which are still insufficient for excitation of significant electroluminescence.

From what has been said, it is easy to understand why, up till now, there have been no observations of electroluminescence from ionic crystals, whereas electroluminescence can readily be observed with the elementary semiconductors Ge and Si (covalent bonds), and it has been observed, although only with relatively

TABLE 1

Compound	Approx. % ionic character of bond	Electroluminescence *	Literature
Ge Si Diamond	0	+	[13] [14, 15] [32]
SiC GaSb InSb	10	+	[16, 27, 38] [39] [17]
GaAs GaP NB InP	20	+	[39] [40] [41] [39]
(Zn, Cd, Hg) (Te, Se, S)	30	+	[2, 23, 42, 43]
AlN	40	+	[44]
CaS, BaS SrS		+	[45, 46]
ZnO	50	+	[47]
Alkali metal iodides†	40-50		
Alkali metal bromides	60	–	
Alkali metal chlorides	70	–	
Phosphor-type oxides of alkaline earth elements	70-80	–	
Alkali metal fluorides	90	–	

* The plus sign denotes electroluminescence and the minus sign denotes
no electroluminescence.

† Of the iodides, only CsI shows electroluminescence [48].

strong fields, in the case of diamond [32], which has the widest forbidden band among covalent crystals. Its forbidden band is approximately 6 eV [33]. In Table 1 we have compared the percentage ionic character of the bonds of various simple binary compounds with the appearance in them of electroluminescence [34]. The estimates of the percentage ionic character of the bonds are based on the electron affinities of the corresponding elements, i.e., on the energies of attraction for bonding electrons by the atoms of the elements, and are only rough approximations. Details of these estimates are given in [35, 36]. The table includes substances from which no electroluminescence has yet been observed, although they luminesce well with other forms of excitation.

It is clear from the table that no electroluminescence has been observed with most ionic crystals. It should also be noted that the percentage ionic character of the bond depends on the mutual polarizability of the ions as well as on the electron affinities. This polarizability reduces the percentage ionic character of the bond; it is most pronounced in compounds formed from elements in the middle groups of the periodic system, since these have more electrons in unfilled shells. This circumstance provides an even greater distinction between substances in the first half of Table 1 and those in the second half.

The literature also contains references to the electroluminescence of other more complex compounds, which cannot at present be classified as in Table 1 because they each contain several bonds which differ from each other in polarity. These compounds are Cu_2O [49]; Al_2O_3 [50]; ZnF_2 [39]; Zn_2SiO_4 [51, 52]; $Cd_2B_2O_5$, $Zn_3(PO_4)_2$, $Cd_3(PO_4)_2 \cdot CdCl_2$, $Ca_2SiO_4 \cdot Mg_2SiO_4$, $Ca_3(PO_4)_2$ [51], and also the above-mentioned ferroelectrics.

In order to produce electroluminescence from compounds with a high percentage of ionic character in their bonds, it is necessary to prepare thin and uniform layers. The reason for this is that, with a reduction in the layer thickness, the conditions for cooling are improved and the development of cascade breakdown is made more difficult. Because of this, the field can be increased. Hence there is an increase in the probability of tunneling ionization of radiation centers, and thus the proportion of the electrical energy which goes into ionization of radiation centers is increased, while the proportion which goes to heat the crystalline lattice is decreased.

For this purpose we prepared a sublimed film of CsI-Tl, about 1 μ thick [48]. This gave an appreciable electroluminescence at a frequency of 20 kcps and a voltage of about 100 V.

These CsI-Tl layers were evidently thin enough to prevent the development of cascade breakdown. In fact, they electroluminesced without breakdown in fields of approximately 2×10^6 V/cm, whereas cascade breakdown of thicker CsI-Tl layers occurred even with fields of approximately 3×10^5 V/cm [53].

In our opinion it should be possible, in this way, to excite electroluminescence in other ionic compounds. However, this would require stronger fields and thinner films, which obviously increases the experimental difficulty. Thus, the best conditions for excitation by an electric field require the use of crystals having bonds of low-percentage ionic character. However, for a crystalline body to be a good electroluminophor, a high probability of radiative transitions is necessary. It is in this connection that there is a relation between the capacities of a crystal for electro- and other forms of luminescence. It is therefore not surprising that the most effective luminophors are those based on compounds of the type $A_{II}B_{VI}$, such as synthetic ZnS electroluminophors with a light yield of about 4% [54]. In these luminophors there is a high probability of radiative transitions combined with bonds of low-percentage ionic character, whereas in the elementary semiconductors Ge and Si, although there are no ionic bonds, there is a low probability of radiative transitions. Thus their effectiveness for electroluminescence is very much less than with ZnS.

CHAPTER II

THE ELECTROLUMINESCENCE OF ZINC SULFIDE

§ 1. Electroluminophors

Electroluminescent samples of zinc sulfide have now been synthesized in the form of finely crystalline powders [55-58], of particle size about 10 μ, and in the form of monocrystals about 1 cm long [59]. Electroluminescent condensers are usually filled with a mixture of finely crystalline powder and dielectric. This obviously increases the electrical strength of the electroluminescent condenser and makes it possible to prepare bright and uniform light sources. Much attention has been devoted to the investigation of thin films, a few microns thick, prepared by distillation of the material in vacuum [60-64]. With such films it is easy to carry out luminescent measurements, both absorptive and electrical.

The spectrum of a zinc sulfide luminophor can be altered over wide limits by varying the activator and coactivator and their concentrations [42, 65]. A ZnS-Cu electroluminophor with a coactivator from the third group (aluminum, gallium, or indium) or from the seventh group (chlorine, bromine, or iodine) gives a green-blue electroluminescence. Addition of manganese gives rise to a yellow radiation [28, 66-71]. Electroluminophors containing manganese have interesting properties. In the first place the yellow radiation evidently does not arise from recombinations and, secondly, some of them radiate brightly under the influence of a constant voltage, even when mixed with dielectric. The brightness is then proportional to the current passing through the film.

There are references in the literature to electroluminophors based on ZnS activated by phosphorus [51] and silver [51, 72].

Studies have also been made of the electroluminescence of mixed crystals in which either some of the zinc has been replaced by another element of the second group (cadmium, mercury [43]) or some of the sulfur has been replaced by another element of the sixth group (selenium, tellurium [42]). Compounds of the type (Zn, Cd, Hg) (S, Se, Te) have the same structure and almost the same lattice constants, so they can form solid solutions in almost any proportions, and the widths of the forbidden bands can alter over a very wide range. In this way is is possible to obtain luminophors giving radiation of any color from the blue to the infrared. The best results are obtained when the zinc is partially replaced by mercury, since this substitution does not reduce the quantum yield significantly and, at the same time, gives luminophors which operate over a wide frequency range for the exciting voltage.

The synthesis of good electroluminophors requires the introduction of more impurities than are needed for the synthesis of photoluminophors. This process is usually accompanied by formation, on the crystallite surfaces, of a phase with a relatively high conductivity, such as Cu_2S, ZnO, etc. Some workers [28, 51] prefer to deposit cuprous sulfide on nonelectroluminescing zinc sulfide, while others [52] mix the nonelectroluminescing phosphor with powdered Cu_2S or ZnO. In either case, the powder acquires the capacity to electroluminesce. Some claim [28, 73] that removing the surface layer from a luminophor by etching reduces its sensitivity to electric excitation, but others [74] have observed no change in the electroluminescent properties of a luminophor when its surface has been etched; it is possible that the result depends on the method of etching. Most authors explain the role of the second phase in terms of concentration of the electric field in the phosphor. In our view, the second phase may also serve as a source of electrons, which penetrate the zinc sulfide and increase its conductivity.

Several workers have stated [42, 43, 72] that the capacity of a zinc sulfide phosphor to electroluminesce is associated with the cubic modification, and that technological changes, resulting in the appearance of and an increase in the quantity of the hexagonal component, lead to a diminution and even to the disappearance of this capacity.

It has been shown [75] that the introduction of stacking disorders into crystals creates the capacity for electroluminescence.

It is now generally believed that the centers which emit light in the course of photoluminescence are located more or less uniformly over the whole of a crystal phophor and are not associated with coarse defects in the crystal lattice such as internal cracks, dislocations, etc. Comparison of the electro- and photoluminescence spectra of a ZnS-Cu, Nd luminophor [76] has shown that the positions and relative intensities of the neodymium lines, which are known to be strongly dependent on the structure of the environment of the activator ions, are similar with both forms of excitation. This means that most of the centers participating in the electroluminescence are located in an undisturbed part of the crystal.

However, crystallites do not electroluminesce uniformly over their whole volume. There are patches which are brighter by several orders of magnitude than other parts. This is probably due to nonuniform distribution of the electric field in the crystal. In the bright parts of the crystal it is clear that the field intensity must be greater than the mean field by at least one order of magnitude, since a field of such intensity is required for the occurrence of the processes described in Chapter I. This field concentration is probably related to the absence on any effect on electroluminescence by magnetic fields up to 10^5 oersteds [77].

It has been shown [76] that with a ZnS-Cu, Er, Mn luminophor the ratio between the manganese band and the erbium lines in the spectrum depends on the excitation density. By comparing the luminances for photo- and electroexcitation for cases when the above ratio was the same, it was found that the effective radiating volume was about 7%. This estimate is only approximate because, as will be seen later, there is also a weak radiation from those parts of the crystal where the field is not high and there are no fast electrons capable of ionizing the crystal lattice and radiation centers.

A very important problem, particularly so in practice, is to explain the nature of the deterioration in luminophors during use.

An empirical formula has been given [78] that relates the decrease in luminance of the electroluminescence B with the time of use t:

$$B = B_0 \left[1 + \frac{t}{t_c} \left(\frac{U_w}{U_m} \right)^2 \right]^{-1},$$

where B_0 is the initial luminance; $t_c \simeq 1000$ hr; U_w is the working voltage; U_m is the voltage at which the luminance is measured periodically.

It has been suggested [78] that this deterioration is associated with a decrease in the number of activator centers due to the mobility of the ions. However, it has been shown [79] that the change agrees better quantitatively with the change in the number of centers responsible for field concentration in the phosphor rather than with the change in number of radiative centers. It has been suggested [80] that the ageing of an electroluminophor is associated with the action of atmospheric oxygen. On the other hand, it has been reported [5] that the life of an electroluminophor can be increased by a factor of 100 or 1000 by means of a waterproof film.

§ 2. The Technique of Measurement

The Electroluminescing Condenser. We have investigated electroluminophors of the composition ZnS-Cu, Al (Cu: 5×10^{-4} to 10^{-3} g-atom/g-mole, and Al: 10^{-4} to 2×10^{-3} g-atom/g-mole). Two types of electroluminescent condenser were prepared from these powders, having different forms of dielectric. A liquid

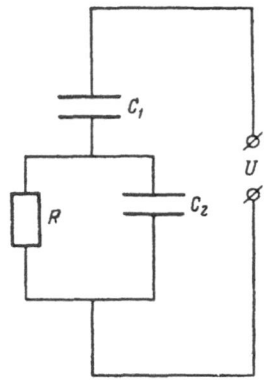

Fig. 2. The equivalent circuit of an electroluminescent condenser.

dielectric, normally three volumes of melamine-formaldehyde resin to one volume of phosphor powder, was used in the first type of condenser. The suspension was applied to a conducting glass, which formed one electrode of the condenser, and a metallic electrode was applied from above. The layer was of constant thickness, about 0.2 mm.

In condensers of the second type the dielectric was a polymerizing mixture of two resins, melamine-formaldehyde and alkyd resins. The condenser was prepared as follows. One volume of powder was treated with one volume of benzene and two volumes of the resin mixture (one part of alkyd resin to two parts of melamine-formaldehyde). The product was well mixed for 4-6 hr in a quartz ball mill with metallic balls, and then the suspension was sprayed onto glass with a conducting coating using a pulverizer of special design. The film was dried for 15 min at room temperature, and then for 4-6 hr in an oven at 373°K. The film thickness was 0.03-0.10 mm. The second electrode of aluminum was applied by vacuum sputtering.

The dielectric effectively insulated the electroluminophor from the electrodes. This was confirmed by the following facts: first, the current passing through the condenser was purely capacitative in nature; second, insertion of mica between an electrode and the electroluminescent layer led to changes which could be compensated for by a simple increase in voltage.

For the simplest equivalent circuit, an electroluminescent condenser can be treated as made up of two layers, the dielectric and the phosphor in series, each with its own capacity and resistance. The dielectric used showed very little loss, and this could be neglected. The equivalent circuit was, therefore, as shown in Fig. 2, where C_1 is the capacity of the dielectric layer between the crystallites of the phosphor, R* and C_2 are the resistance and capacity of the phosphor layer, and U is the exciting voltage.

We will carry out an approximate calculation of the voltage distribution for a system corresponding to the circuit of Fig. 2. When an alternating voltage of amplitude U_0 is applied to such a system, the voltage drop across the capacity C_1 will be

$$U_{c_1} = \frac{U_0 \left(\frac{1}{i \omega C_2 R} + 1 \right)}{1 + \frac{C_1}{C_2} + \frac{1}{i \omega C_2 R}} \, . \qquad (\text{II.1})$$

The parameter Z may be introduced into this equation, where

$$Z = \omega C_2 R. \qquad (\text{II.2})$$

At low frequencies, for which

$$Z \ll 1, \qquad (\text{II.3})$$

$$U_{c_1} \to U_0, \qquad (\text{II.1'})$$

i.e., practically all the voltage is concentrated across C_1. At high frequencies, for which

$$Z \gg 1, \qquad (\text{II.4})$$

$$U_{c_1} \to \frac{U_0}{1 + \frac{C_1}{C_2}}, \qquad (\text{II.1''})$$

_____

* R would be a nonlinear resistance, but this need not be taken into account in an approximate calculation.

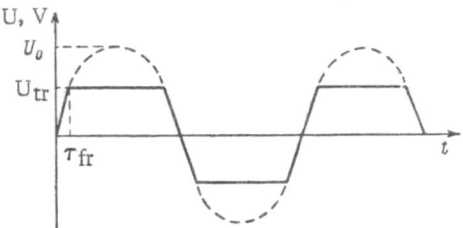

Fig. 3. Oscillograms of sinusoidal voltage (dotted line) and of the same voltage after passing through the amplitude restrictor (continuous line). τ_{fr} is the duration of the front of the voltage pulses.

and the voltage across the resistance R corresponding to the phosphor layer is

$$U_R \to \frac{U_0}{1 + \dfrac{C_2}{C_1}} . \qquad (II.5)$$

It is clear from equation (II.5) that, at frequencies for which condition (II.4) is fulfilled, the voltage applied to the phosphor layer is independent of the frequency and can be expressed by a real value, i.e., it is in phase with the full voltage applied to the system. This condition was evidently fulfilled for our electroluminescent condensers. In fact, appreciable electroluminescence was observed even at frequencies of about 1 cps. Thus, a sufficient voltage was being applied to the phosphor layer even at this frequency. In other words equation (II.1') did not apply at this frequency, i. e., condition (II.3) was not valid. This means that $Z \gtrless 1$. Most of our measurements were carried out at a frequency $f \geq 50$ cps, so that condition (II.4) was evidently well fulfilled in our experiments. We will therefore assume in further calculations that the voltage across the phosphor layer is in phase with the total voltage applied to the electroluminescent condenser.

The capacity of the dielectric film of the electroluminescent layer was obviously of the same order, and possibly greater, than the capacity of the electroluminescent crystallites. Moreover, considerable electroluminescence was observed when 10 V overall was applied to an electroluminescent layer of thickness approximately equal to the size of one crystallite. In this case the voltage acting across the phosphor layer must have been at least about 2.5 V, since light quanta with an energy of about 2.5 eV were emitted. It therefore follows from equation (II.5) that

$$\frac{C_2}{C_1} \lesssim 3 . \qquad (II.6)$$

It is clear from equation (II.5) that the voltage acting across the phosphor layer depends on the capacity of the dielectric film, which is proportional to the dielectric constant ε. Let us consider how this voltage alters with a change in the capacity of the dielectric. According to equation (II.5), a change in the capacity C_1 by a factor k_1 should lead to a change in the proportion of the voltage acting across the phosphor by a factor

$$k_2 = \frac{\dfrac{C_2}{C_1}(k_1 - 1)}{\left(\dfrac{C_2}{C_1} + 1\right)\left(\dfrac{C_2}{C_1} + k_1\right)} . \qquad (II.7)$$

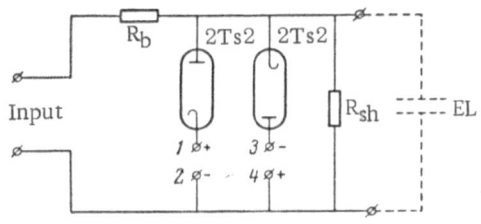

Fig. 4. Block diagram of amplitude restrictor. EL is the electroluminescent condenser.

Excitation and Registration of Electroluminescence. To excite the radiation from an electroluminescent condenser, we applied both sinusoidal and trapezoidal forms of voltage.

The source for the sinusoidal voltage was a ZG-10 audio-frequency generator with a step-up transformer on the output. The trapezoidal form of voltage was either constant with time or altering at a uniform speed. This made it possible to separate the various stages of emission more easily.

The trapezoidal voltage was obtained by restricting the amplitude and thus "cutting off" the peaks of the sinusoidal voltage (Fig. 3).

When the sinusoidal voltage was "cut off" at a level equal to half its amplitude, the deviation from linearity of the front of the resulting trapezoidal voltage was less than 5%. The amplitude of the trapezoidal voltage could be adjusted by changing the "cut off" level. Alterations to the slope of the front and to the frequency of the trapezoidal voltage were carried out, respectively, by varying the amplitude and frequency of the sinusoidal voltage. Figure 4 is a block diagram of the amplitude restrictor.

A high voltage, sinusoidal in form, was applied to the input terminals. A constant voltage, whose value was determined by the "cut off" level for the sinusoidal voltage, was applied to terminals 1, 2 and 3, 4. R_b was a blocking resistance to restrict the current to a value suitable for the diodes and to avoid breakdown of the electroluminescent condenser. The value of the shunting resistance R_{sh} was selected so as to be considerably less than the capacitive resistance of the electroluminescent condenser at the frequencies used; its function was to eliminate the effect of the latter on the wave form.

The luminance of the electroluminescent condenser was measured with a specially built apparatus, using an FÉU-19 photomultiplier as the light-sensitive element. Figure 5 shows a block diagram of the apparatus, where 1 is the electroluminescent condenser; 2 is a color filter; 3 is an FÉU-19 photomultiplier; 4 is a differential cathode-follower unit; 5 is the generator for the exciting voltage; 6 is the high-voltage source for the FÉU-19; 7 is the measuring instrument.

The measuring instrument for medium luminance was a PM-70 milliammeter (10 mA, internal resistance 70 ohms). The instantaneous luminance was investigated with a double-beam Cossor oscillograph, the second channel of which was controlled by the exciting voltage.

The use of a differential cathode follower made it possible to measure luminances from 10^{-5} apostilb upwards. The instrument was calibrated and controlled by comparison with standard constant light sources, which had been photometered with a GOI absolute photometer. Figure 6 shows the circuit of the differential cathode-follower unit. The sensitivity of the instrument was adjusted by switching in the resistances R_g.

In some cases we made use of an MPO-2 eight-loop oscillograph. It was then possible to obtain oscillograms of several processes at once, and also to photograph oscillograms so as to be able to record the history of a process over a long time interval. It was in this way that we recorded the build-up of electroluminescence. For these experiments we constructed a quite simple direct current amplifier, whose circuit is shown in Fig. 7. Using this amplifier it was possible to work with the highest-frequency vibrator applicable to the MPO-2 oscillograph (up to 10 kcps), requiring amplification of the signal to a current of about 100 mA.

The first stage of the amplifier, consisting of 6Zh3P tubes, amplified the signal voltage from the photomultiplier, and the second stage, consisting of 6P3 tubes connected to give a differential cathode follower, amplified

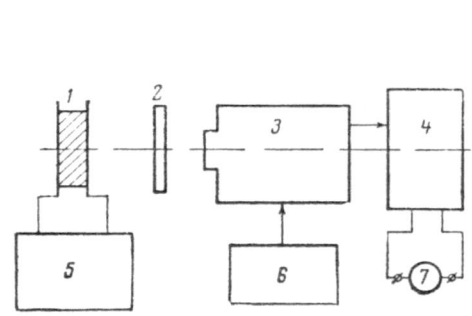

Fig. 5. Block diagram of photometric apparatus.

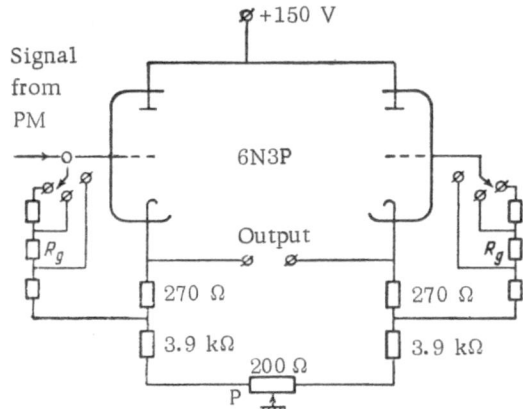

Fig. 6. Circuit of differential cathode follower.

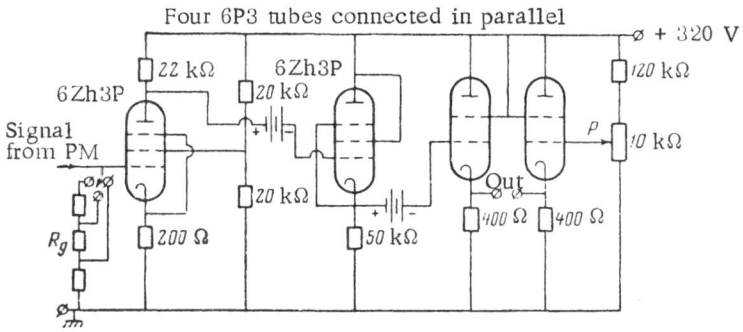

Four 6P3 tubes connected in parallel

Fig. 7. Circuit of direct current amplifier.

the signal current. So that the output stage should not shunt the anode load of the voltage amplifier, a cathode follower composed of 6Zh3P tubes was connected between them. The resistance R_g could be altered to change the sensitivity, and the potentiometer P could be adjusted to set the zero in the absence of a signal.

The condenser could be placed in a thermostat (Fig. 8) when it was necessary to investigate electroluminescence at various temperatures. In the figure, 1 is a quartz Dewar vessel; 2 is the lid of this vessel, and through it slides the textolite tube 3, joined to a metal vessel 4; 6 is a metal plate which provides thermal contact between one electrode of the electroluminescent condenser 7 and the electric heater 5; 8 is a thermocouple mounted inside the plate 6. The temperature of the condenser could be altered by changing the electric voltage feeding the heater and by varying the distance between the electroluminescent condenser and the metal vessel 4, filled with liquid nitrogen. A film of vacuum grease was placed between the lid 2 and the Dewar vessel 1, and between the lid and the tube 3, in order to prevent atmospheric moisture from getting into the Dewar vessel.

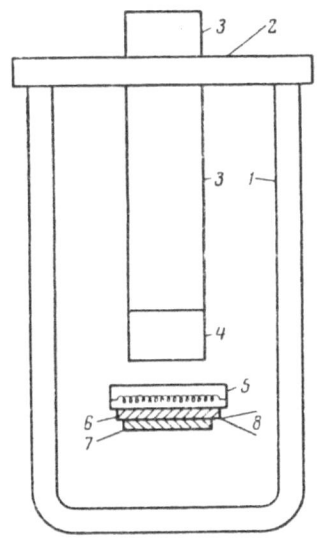

Fig. 8. Diagram of thermostat.

Measurement of the Electrical Energy Consumed in Electroluminescence. The power oscillator of a loop oscillograph was used to measure the energy consumed in electroluminescence. The loop of this oscillator was fed with a signal proportional to the current through the condenser being investigated, and the coil of the electromagnet was fed with a signal proportional to the voltage. Then the movement of the oscillator spot was proportional to their product, i.e., to the power. However, the sensitivity of the oscillator towards current was very small, and, in order to record the power, it was necessary to feed a current of several milliamperes to the loop. The currents passing through an electroluminescent condenser were less than this by several orders of magnitude; they therefore required amplification. The amplifier used for this purpose was similar to that shown in Fig. 7.

The absence of phase distortion was checked by obtaining oscillograms of the power consumed by a normal resistance and a normal capacity. Measurement of the power absorbed by a known resistance made it possible to calibrate the oscillograms in absolute units. An oscillogram of the power for an electroluminescent condenser, unlike that for an ideal condenser, was asymmetric relative to the zero line (Fig. 9), since energy was absorbed by an active component of its impedance in both half-periods of the voltage. In order to calculate the mean power consumed by an electroluminescent condenser, it was necessary to subtract the smaller half-period figure from the larger and average the result for a period.

19

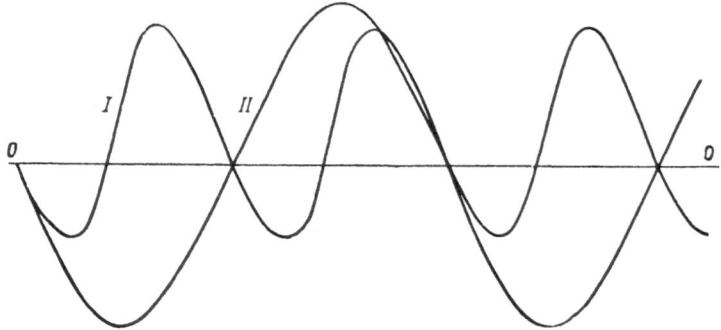

Fig. 9. Oscillograms of (I) power and (II) excitation potential of an electroluminescent condenser. The line 0-0 represents zero values for the power and potential.

In order to make the method more precise, we eliminated the zero mean power corresponding to the oscillations in the energy of the condenser without loss. This was done as follows. A voltage equal, but opposite in phase, to the voltage exciting the electroluminescent condenser was fed to an air condenser. The total current through both these condensers was fed to the oscillator loop. The reactive component on the power oscillogram was then eliminated by varying the capacity of the air condenser. The power absorption wave could then be scaled up to span the whole screen, and the relative error in determining the area of the oscillogram was considerably reduced. This elimination of the capacity component of the power could be carried out regardless of the form of the active component, so that nonlinearity of the electroluminescent condenser could be allowed for completely.

§ 3. The Oscillogram of the Luminance of an Electroluminescent Condenser

When a constant voltage is applied to a mixture of phosphor and dielectric, the phosphor crystals are gradually polarized and the electric field is transferred to the dielectric. It is therefore necessary to use an alternating electric field, and it is impossible to achieve a stationary state in the strict sense of the word. Even if a constant mean luminance is established, the instantaneous value oscillates between certain limits at a frequency equal to twice that of the applied voltage. This doubling is due to the fact that the crystal regions in which the radiation is excited are oriented in various directions in an electroluminescent condenser, so that with one polarity of the applied voltage the field is concentrated in one set of regions, and with the opposite polarity the field is concentrated in other regions. The total radiation from both sets of regions therefore oscillates with half the period of the voltage. In fact, investigations of the luminescence of crystallites under the microscope showed that each individual radiating patch emitted light with a period equal to that of the voltage [28, 81, 82].

Figure 10 shows an oscillogram of the luminance of an electroluminescent condenser together with an oscillogram of the exciting sinusoidal voltage.

The luminance oscillogram consists of a variable component, usually called the wave luminance, and a constant component, denoted by K in the figure. In each half period of the exciting voltage the wave luminance has a main peak (denoted by A). Under certain conditions some phosphors show an additional peak B, which is normally smaller than the main peak. The components of the wave luminance are better resolved when the excitation voltage is trapezoidal in form (Fig. 11). We may note the following special features of the wave luminance of electroluminescence excited by a trapezoidal voltage whose mean value corresponds to a mean field in the electroluminescent condenser of less than 10^5 V/cm:* (a) sharp maximum at the instant of

---

* The form of the wave luminance excited by a higher voltage is discussed below.

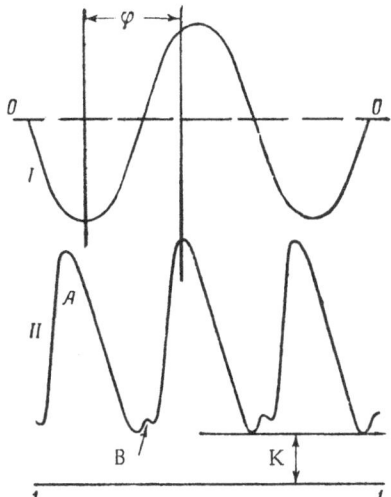

Fig. 10. Oscillograms of (I) the exciting voltage and (II) the luminance of an electroluminescent condenser. 0-0 is the zero voltage line and 1-1 is the zero luminance line; φ is the phase angle between the start of voltage reversal and the maximum of the main luminance peak.

completion of the voltage rise, * and (b) a drop in electroluminescence when the voltage becomes constant.

We investigated the relation between the form of the wave luminance and the dielectric used. It was found that a change from one dielectric to another only affected the phase of the auxiliary peak in the wave luminance; this was associated with differences in the dielectric constants of the dielectrics and could be compensated for by changing the voltage applied to the condenser. The following dielectrics were investigated: melamine-formaldehyde resin, castor oil, silicone oil, condenser oil, calorie 2.

The occurrence of wave luminance was clearly associated with movement of electrons in the crystal, which was regulated to some extent by the electric field.

It is now considered as established [28, 39, 83, 84] that, with a ZnS-Cu electroluminophor, in each half period of the voltage there occurs ionization of the radiation centers at the cathodic edge of the crystal and movement of electrons to the other edge of the crystal. In the succeeding half period the electrons are returned to the ionized centers and recombine with them, giving off radiation.

This two-stage model of the electroluminescence cycle is confirmed by the following facts:

1. There is very small probability of recombination in the region of the strong field where ionization of radiation centers occurs, because, with an electron mobility in ZnS of 120 cm^2/ V·sec [28] and a crystal size of about 10^{-3} cm, the electrons move out rapidly and are unable to recombine with the ionized centers of the excitation region, which are smaller than the crystals by at least one and possibly two orders of magnitude.

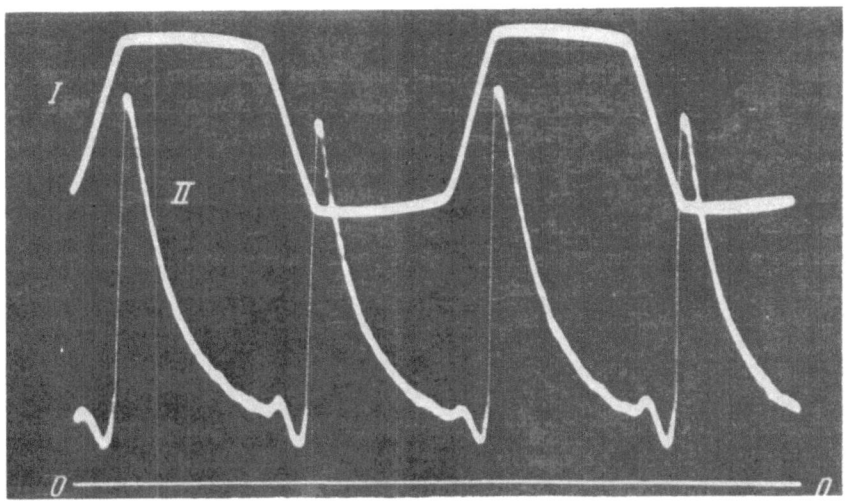

Fig. 11. Oscillogram of (I) trapezoidal exciting voltage and (II) the luminance of an electroluminescent condenser. 0-0 is the zero luminance line.

---

* A corresponding maximum of the wave luminance at the instant of cessation of the increase in voltage was confirmed by us when the frequency of the exciting voltage was varied from 20 to 10,000 cps.

TABLE 2

T, °K	φ/π	
	200 cps	5500 cps
77	0.74	0.78
293	0.79	0.86
373	0.81	0.89

2. When a voltage is applied to an electroluminescent condenser containing a strongly luminescing phosphor, significant radiation is only observed when the voltage is switched off or it polarity is reversed (see Fig. 25a), i.e., when the electrons can return to the excitation region.

3. Superposition of a weak single pulse onto a sinusoidal exciting voltage has its main effect, not on the existing peak, but on the subsequent peak of the luminance wave [28, 85].

This fact can again be explained by supposing that electrons can move back to the radiation centers only after the voltage is reversed, i.e., in the next half period.

It is known from investigations on the photoluminescence of ZnS-Mn that its radiation includes a nonrecombination part. This is derived from luminescence centers, which are not ionized but only excited, and evidently explains some special features of the electroluminescence of such phosphors. In particular, their radiation occurs in the same half period as their excitation, since the field evidently has little effect on the emission from such centers.

The variation with temperature and frequency of the phase of the auxiliary peaks has been investigated for ZnS-Cu luminophors [84, 86]. It was shown that these peaks are associated with the return to the excitation region of electrons freed by thermal ionization from entrapment centers of depth less than 0.4 eV. It has been asserted [87] that the main peaks are also due to the return of electrons freed by thermal ionization of the traps. However, this is contradicted by the weak dependence of their phase on the temperature and on the frequency of the exciting voltage.

According to our observations, the phase φ of the maximum of the main peak of the luminance wave (Fig. 10) varies little with a change in the temperature from 77 to 373°K, or with a change in the frequency f of the exciting sinusoidal voltage from 200 to 5500 cps (Table 2).

The phase φ is refered to the instant of the start of reversal of the applied voltage, i.e., to the instant when the electrons can return to the excitation region.

If the main peaks are produced by return to the radiation centers of electrons freed from traps by thermal ionization, then their phases should depend on the frequency and very much so on the temperature. In fact the probability of thermal ionization is proportional to the time allowed for it (in this case the period of the voltage) and increases exponentially with temperature. The behavior of the auxiliary peak in the wave luminance corresponds precisely to this change in the probability of thermal ionization [84, 86].

Some authors have suggested [28, 84] that the main peaks are formed by return of free electrons to the excitation region. However, considering that the mobility of electrons in ZnS is about 120 cm^2/V·sec [28], it will be appreciated that the time for return of free electrons is very small compared with the duration of a half period of an audio-frequency voltage. In other words, if it is supposed that the main peak is formed by return of free electrons to ionized centers, it would be expected to occur close to the voltage reversal point. However, experiment shows that the main peak has its maximum at the instant when the growth in voltage is complete (Fig. 11).

It is evident from an oscillogram of the radiation from an electroluminescent condenser excited by an alternating voltage (Figs. 10, 11) that, in addition to the variable component, there is a constant background, denoted by K in Fig. 10. When this constant component is mentioned in the literature, it is usually assumed that it is formed by overlapping of the wave luminance peaks. In other words, a subsequent burst of radiation starts as the previous one dies away. However, it will be shown in §7 that this is not the case.

§ 4. The Main Peaks of the Electroluminescent Luminance Wave

We have already stated that the maximum of the luminance wave of electroluminescence excited by a trapezoidal voltage corresponding to a mean field in the condenser of less than about 10^5 V/cm coincides with the instant of completion of the growth in voltage. This applies over a wide frequency range.

TABLE 3

f = 265 cps		f = 1325 cps	
U_0, V	φ/π	U_0, V	φ/π
50	0.89	80	0.96
75	0.87	140	0.93
130	0.83	220	0.9
315	0.78	355	0.87
400	0.75	560	0.82
530	0.71	670	9.78

Note: T = 350°K.

In the case of excitation by a sinusoidal voltage, the maximum of the main peak is located near the instant of slowing up of the growth in voltage. Here the phase angle (Fig. 10) depends on the value of the voltage (Table 3).

These facts show that the electric field plays an essential part in the formation of the main peaks. However, we have already noted that the action of the field cannot be to cause a simple drifting transfer of free electrons. We therefore assumed [88] that the main peak of the luminance wave must be produced by return to ionized radiation centers of electrons which had been held by traps at the opposite edge of the crystal during the preceding half period. These electrons are freed from the traps by the electric field, either directly by tunneling penetration of electrons or by impact ionization. In other words, when an alternating voltage is applied to an electroluminescent condenser, it not only ionizes the crystalline lattice and radiation centers, i.e., creates a light sum, but there is also a realization of the light sums which were created in the previous half period of the voltage.

It should be possible to check our assumption by the following experiment. A voltage is applied to an electroluminescent condenser for some period of time. The voltage is then switched off, and the condenser is irradiated with infrared light. The latter causes emission of the light sum stored in the condenser, i.e., there is a pulse of luminescence. If the light sum in electroluminescence is in fact emitted mainly at the expense of the field, then the magnitude of this pulse should depend very little on the time interval between switching off the voltage applied to the condenser and irradiation by infrared light.

We carried out such an experiment [89] in the following way. An electroluminescent condenser excited by voltage pulses of alternating polarity was for a short time irradiated by a pulse of infrared light, isolated by an RG-9 spectral filter, 4 mm thick, from the spectrum of an IFK-500 pulse lamp. The duration of the voltage pulse τ_1 and the time between pulses τ_2 were both 5 msec. The amplitude of a voltage pulse, U_0, was 170 V. The duration of the infrared pulse was 0.8 msec. The approximate spectral composition of the light pulse was calculated in accordance with published data [90] and is shown in Fig. 12. Figure 13 shows the oscillograms of this experiment: trace 1 is that of the exciting voltage; trace 2 is that of the infrared pulse; trace 3 is that of the electroluminescent emission. The main peak of the luminance wave before irradiation of the condenser by infrared light is denoted by A, the first peak after irradiation by A_1, the second by A_2, and so on. The peak of the infrared irradiation is denoted by I.

Figure 13a shows an oscillogram of the luminance for the case when the pulse of infrared light occurs between voltage pulses. G denotes the burst of radiation induced by the infrared light. Figure 13b shows an oscillogram of the luminance for the case when the pulse of infrared light occurs during a voltage pulse; it is clear from the figure that the infrared light causes a marked reduction in the radiation from the electroluminescent condenser. The different effects of the infrared light in the two cases can evidently be explained as follows. Application of the voltage causes significant liberation of electrons by the field itself, without the infrared light. Irradiation by infrared light at this instant cannot, therefore, cause any significant increase in the concentration of free electrons; it may, indeed, weaken the emission to some extent, owing to its quenching action (liberation of holes). However, if the irradiation by infrared light occurs at an instant when the voltage is not applied, then it will considerably increase the concentration of free electrons, and its quenching action will be masked by a pulse. It is clear from Fig. 13 that the light sum for the peaks G and A_1 is considerably less than that for the peak A

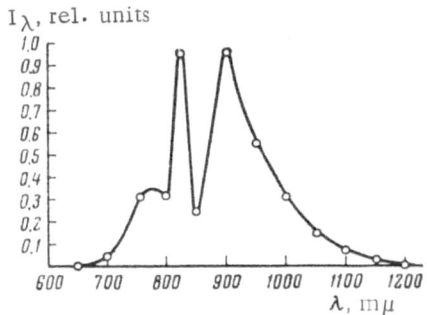

I_λ, rel. units

Fig. 12. Spectral composition of light pulse.

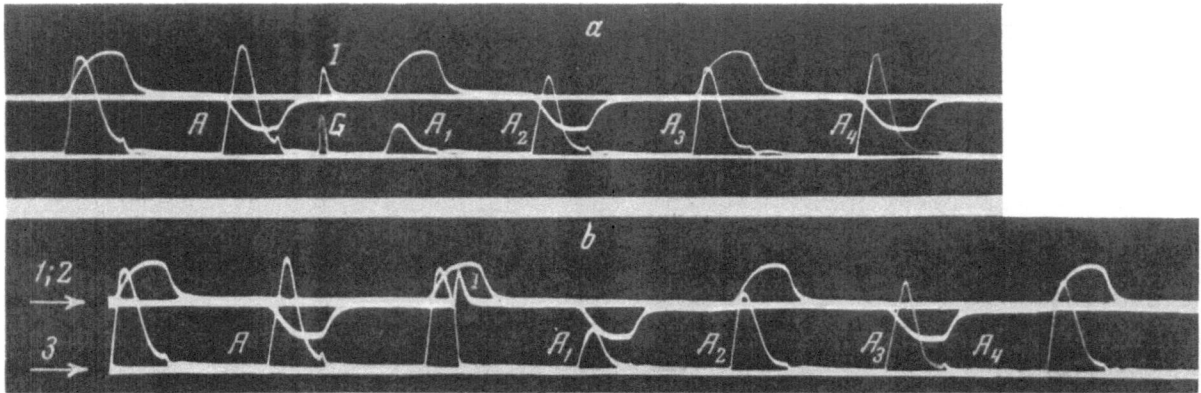

Fig. 13. Oscillograms of (1) excitation voltage, (2) intensity of pulse lamp radiation, and (3) luminance of electroluminescent condenser.

preceding the infrared pulse. This reduction in light sum may be associated with a return of the electroluminescent condenser towards its equilibrium state under the action of the infrared light. Indeed, it is clear from Fig. 13 that a build-up of electroluminescence (successive increases in the luminance wave) begins after the effect of the infrared pulse. This is similar to the build-up in emission observed with a previously irradiated electroluminescent condenser (see Fig. 25).

As we expected, the value of the luminescence pulse produced by the infrared light (peak G in Fig. 13a) varied very little with the time interval between the instants of switching off the voltage and of infrared irradiation. At room temperature (Fig. 14), it remained practically unaltered over a time interval equal to that between the voltage pulses ($\tau_2 = 5$ msec). The role of thermal liberation of electrons and holes increased with increasing temperature. This evidently explains the reduction in size of this pulse, for the same time conditions, in an experiment where the electroluminescent condenser was heated to 433 °K (Fig. 15).

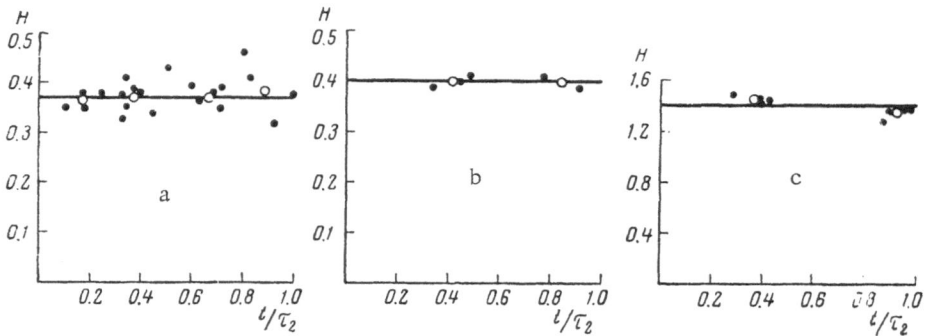

Fig. 14. Variation in amplitude of the peak G produced by the infrared light with the time since the voltage was shut off (T = 293°K): a) for the whole, b) for the green, and c) for the blue emission from the electroluminescent condenser. H is the ratio of the amplitudes of peaks G and A; t is the time since the voltage was shut off and τ_2 is the time interval between voltage pulses; ● are experimental points; ○ are averaged results from a series of experimental points.

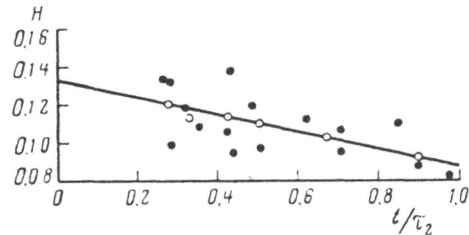

Fig. 15. The symbols have the same meaning
as in Fig. 14a, but T = 433°K.

The luminescence pulse (peak G) achieved its maximum some time t_m after switching on the infrared light. Making use of the value of t_m, and knowing the infrared light intensity, we could estimate the strength of the oscillator corresponding to the liberation of electrons and holes from a localized level under the influence of the infrared light. This calculation is carried out in Appendix I. The value obtained is about 5×10^{-3}.

The luminophors which we investigated had two emission bands, green and blue. Under the excitation conditions used, the green band was more intense by an order of magnitude than the blue band. The base level for a "blue" center is closer to the valence band than is that for a "green" center [91, 92], so that, under normal conditions, the transfer of holes from "blue" to "green" centers exceeds the transfer in the opposite direction. Irradiation by infrared light evens up the probabilities of liberation from "blue" and "green" centers, and thus alters the balance in favor of transitions from "green" to "blue" centers. Since the green band is more intense than the blue by an order of magnitude, relatively more holes can transfer to the blue centers. This evidently explains why H, the ratio of the amplitude of peak G to the amplitude of peak A, was about 0.4 for the green band and 3.5 times as much for the blue band.

Another difference has been noted in the behavior of the blue and green electroluminescence bands [93]. When the electroluminescence is excited by voltage pulses, the luminance of the green band drops rapidly at the instant of switching off the voltage, whereas in the case of the blue band this only occurs if the interval between switching on and switching off the voltage is not less than some critical value of about 30 μsec. * This behavior of the blue band is difficult to understand, since, when the voltage is switched off, there is a rapid reduction in the liberation of electrons from traps and in their uptake by ionized centers.

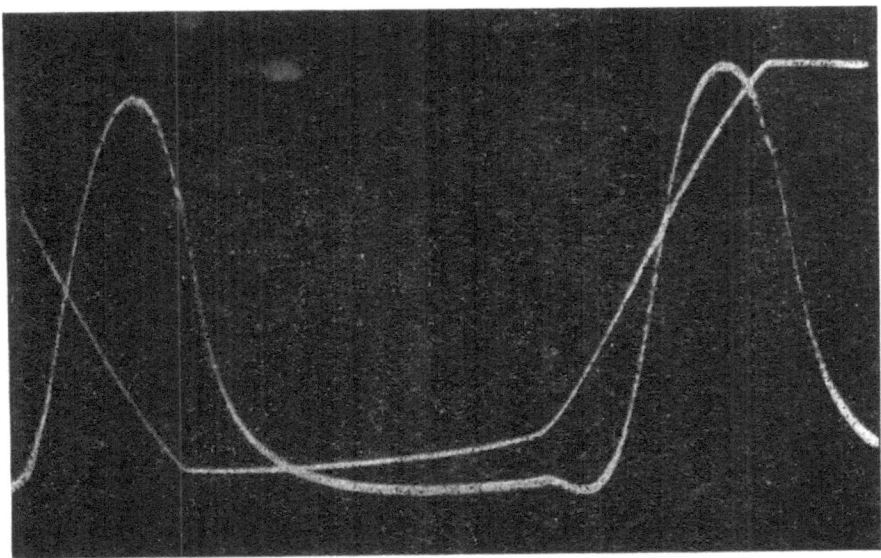

Fig. 16. Oscillogram of the luminance of an electroluminescent condenser excited
by a trapezoidal voltage exceeding the critical.

---

* A difference in the attenuation rates of the blue and green bands has been observed [94], over the same time interval, with excitation by a short-lived spark or by α-particles.

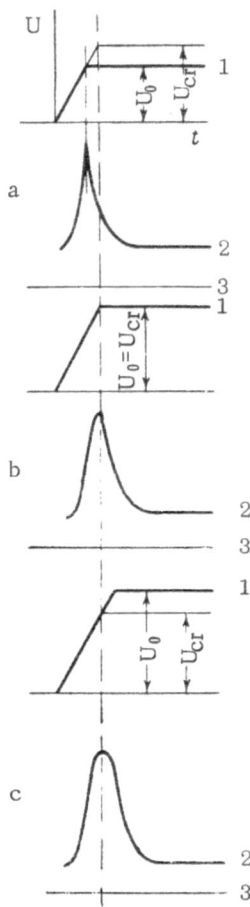

Fig. 17. Dependence of the position of the main peak maximum on the amplitude of the voltage pulse: 1) oscillograms of exciting voltage; 2) oscillograms of luminance; 3) zero luminance level. U_0 and U_{cr} are the amplitude and critical value of the voltage respectively.

This difference in the behavior of the blue and green bands has been explained [95] in terms of the diffusion theory of luminescence* [96]. This theory takes into account the fact that ionized luminescence centers, having excess positive charge, must attract electrons. Even at relatively high separations, the potential energy of an electron in the field of such a center will exceed the electron's energy of thermal movement, so that the electron cannot move right away from the center but gradually moves towards it until they recombine. Thus, if a high proportion of the electrons are in spheres of capture by ionized centers, then the radiation will follow a quasi-monomolecular and not a bimolecular law. If the residence time of an electron in the captured state at a blue luminescence center is about 10 μsec, and considerably less at a "green" center, then, in a time interval of one microsecond, the green luminescence will respond rapidly to the concentration of free electrons, but the blue only after a delay. During this time the luminance of the blue band will greatly diminish on its own, so that it will seem from experiment that this band does not react at all to the external field, altering the concentration of free electrons.

We will turn now to another point. If our interpretation of the process of formation of the main peaks is correct, then at a high enough voltage their maximum should no longer correspond with the end of the growth in voltage, but should occur somewhat earlier. Indeed, with increasing electric field, trapping centers at shallow levels should be destroyed first, then deeper centers, and finally, at some value of the electric field which we will call the critical value, there should be almost complete removal from even the deepest capture centers. As the result of depletion of the local levels, acting as electron sources, a further increase in the electric field should be accompanied by a reduction in luminance. This effect has, in fact, been observed [88] at voltages corresponding to a mean field in the electroluminescent condenser of more than a limiting value approximately equal to 10^5 V/cm (Fig. 16).†

In Fig. 17a the amplitude of the exciting voltage pulse U_0 is less than the critical voltage U_{cr}, and the luminance wave achieves its maximum at the end of the front of the voltage pulse (see also Fig. 11). In Fig. 17b the amplitude of the voltage pulse is equal to the critical value, and in Fig. 17c it is greater. In this series the luminance wave passes through a maximum when the exciting voltage pulse reaches the critical value. The critical value for the voltage slowly increases with increasing slope of the voltage pulse front (Fig. 18). This increase is obviously associated with the need to use a higher voltage to liberate electrons in a shorter time.

---

* The same viewpoint was used [95] to explain the differences in behavior of the blue and green bands in electroluminescence and in excitation by a spark and by α-particles.

† It should be emphasized that in this case we are speaking of the form of the wave luminance, i.e., of the distribution in time of the light sum emitted by an electroluminescent condenser over a voltage cycle, and not of the actual value of this light sum.

Thus the form of the main peak of the luminance wave can be described qualitatively on the basis of our concepts as to the process of its formation. However, to make a further check of the correctness of these concepts, it was necessary to carry out quantitative comparisons of theoretical and experimental relations for the variation in form of the luminance wave with temperature and with amplitude and frequency of the exciting voltage.

§ 5. Variation with Temperature of the Critical Voltage for the Main Peaks of the Luminance Wave [88]

Figure 19 shows the variation with temperature of the critical voltage at 200 cps. The critical voltage has a maximum value at about 250°K. It drops continuously at higher temperatures and, at lower temperatures, drops to an approximately constant value at about 80°K.

We investigated how the temperature affected the dielectric constant ε (Fig. 20) of the dielectric used, a mixture of melamine-formaldehyde and alkyd resins, to see whether this could account for change in critical voltage with temperature.

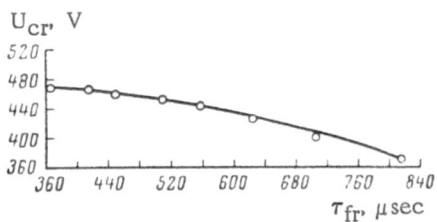

Fig. 18. Dependence of the critical voltage on the duration of the front of the exciting voltage pulse: U_0 = 250 V; f = 200 cps.

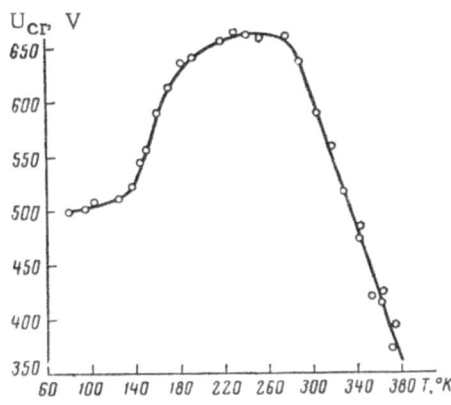

Fig. 19. Variation with temperature of the critical voltage.

We may note first that the curve of Fig. 19 has a maximum while the curve of Fig. 20 shows a continuous increase with rising temperature. Moreover, it follows from equations (II.7) and (II.6) that, with a change in temperature from 140 to 200°K, the critical voltage should decrease by about 1.5% owing to the increase in ε. Figure 19 shows that the critical voltage did not decrease, but increased by 25%. It decreased by 35% when the temperature was increased from 280 to 360°K, whereas, according to equations (II.7) and (II.6) and the change in ε, it should decrease by 6%. Thus the variation with temperature of the critical voltage (Fig. 19) is not the result of the change with temperature of the dielectric constant of the dielectric used.

We have suggested [88] that the variation of critical voltage is a result of the variation with temperature of the probability that the electric field liberates electrons from capture centers, either directly or by impact ionization.

Let us compare the variation with temperature of the critical voltage (Fig. 19) and the variation with temperature of the critical fields* for the processes of impact ionization and of electron liberation by tunneling (field ionization).

The field enters into equation (I.9) for the probability of impact ionization in the form E_{imp}/E. Consequently, the probability reaches a definite value when the ratio E_{imp}/E has a definite value. Thus the critical field E_{cr}^{imp} is proportional to E_{imp}. Therefore E_{cr}^{imp} alters with temperature in the same way as E_{imp}. Thus it follows from (I.8) that

$$E_{cr}^{imp} = E_{cr}^0 \sqrt{\frac{e^{\hbar\omega_0/kT} + 1}{e^{\hbar\omega_0/kT} - 1}}, \qquad (II.8)$$

---

* In theory the critical field for breakdown is the field at which the ionization probability reaches a definite value. We assumed that the field had a critical value if the product of the probability for liberation of an electron from a capture center and the observation time τ_{obs} was equal to unity. The observation time was taken as the duration of the front of a trapezoidal pulse.

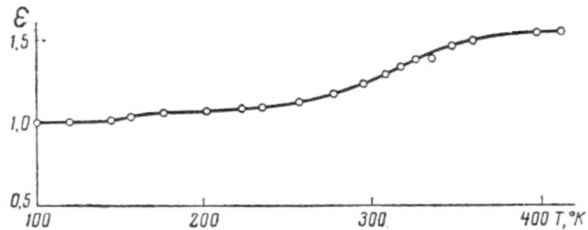

Fig. 20. Variation with temperature of the dielectric constant of the polymerized mixture (1 part of alkyd resin to 2 parts of melamine-formaldehyde) used in the electroluminescent condenser. The dielectric constant has been expressed in terms of its value at 100°K.

where E_{cr}^0 is a value independent of temperature and equal to the critical field at zero temperature. According to this equation, the critical field decreases with falling temperature until kT becomes much less than $\hbar\omega_0$. At this point E_{cr}^{imp} approximates to the constant value E_{cr}^0. This means that the equation corresponds to the low-temperature branch of Fig. 19.

Equation (II.8) may be put in a form suitable for calculating the value of the optical phonon:

$$\hbar\omega_0 = kT \ln \frac{(E_{cr}^{imp})^2 + (E_{cr}^0)^2}{(E_{cr}^{imp})^2 - (E_{cr}^0)^2}. \tag{II.9}$$

Since the probability of impact ionization and direct field ionization (penetration of electrons by tunneling) increase rapidly with increasing electric field, it is clear that the processes of liberating electrons from traps will be most effective in regions where the field is concentrated, i.e., at the edges of a crystal where the field is proportional to the square root of the applied voltage. Thus the voltage plotted on the ordinate axis of Fig. 19 is proportional to the square of the critical field, i.e., to E_{cr}^2. It is clear from Fig. 19 that, already at 80°K, the critical voltage is very little dependent on the temperature. We may therefore take it that $E_{cr}(80°K) \approx E_{cr}^0$. The value of the longitudinal optical phonon calculated from equation (II.9) was 3.8×10^{-2} eV. The value of this phonon obtained from the vibrational structure of the edge emission spectrum and from the Raman spectrum was 4.3×10^{-2} eV [97-99].

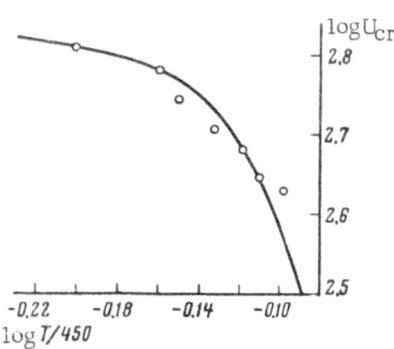

Fig. 21. Variation with temperature of the critical voltage in the region of high-temperature decline. The curve on the diagram is from equation (II.11) for the case when $E \sim \sqrt{U}$.

It is clear from Fig. 19 that at T > 250°K the critical voltage dropped quite rapidly with increasing temperature. In other words, beginning at 250°K an increase in temperature accelerates formation of the main peak of the luminance wave. We have already noted, in Chapter I, that increasing the temperature favors the tunneling penetration of electrons in field ionization of the crystalline lattice and various centers. Thus, with increase in temperature there is a particularly strong increase in the probability of ionization by the combined action of heat and an electric field [see equation (I.5)]. We therefore suggest that at T > 250°K the main peak in the luminance wave is produced by return of electrons, liberated from capture centers for the most part by tunneling transitions, to ionized radiation centers. Such tunneling transitions take place by the simultaneous ionization of capture centers by heat and by the electric field. In the temperature region corresponding to the high-temperature branch of Fig. 19, and with fields less than 10^6 V/cm, equation (I.5) applies for capture centers of depth $\Delta_{c.c}$ greater than 0.4 eV.

The exponential index in equation (I.5) for the critical value of the electric field will be denoted by δ, so that

$$eE_{cr}^{t,T} = \frac{kT\sqrt{2'm\Delta_{c.c}}}{h}\sqrt{1 - \frac{\delta kT}{\Delta_{c.c}}} \; . \tag{II.10}$$

This may be rewritten as

$$\frac{E_{cr}^{t,T}}{E_1} = \frac{T}{T_1}\sqrt{1 - \frac{T}{T_1}}, \tag{II.11}$$

where

$$T_1 = \frac{\Delta_{c.c}}{\delta k}; \quad E_1 = \frac{\sqrt{2'm\Delta_{c.c}^3}}{\delta he} \; . \tag{II.12}$$

In Fig. 21 we compare the experimental relation between critical voltage and temperature (for $T > 250°$K) with the relation given by equation (II.11), assuming that liberation of electrons occurs at the crystal edges, where $E \sim \sqrt{U}$. The graph is drawn with logarithmic coordinates, so as to eliminate the effect of scale factors on the curve shape.

The theoretical and experimental curves of Fig. 21 coincide at $T_1 = 398-403°$. Then according to equation (II.12), the depth of the capture center is given by

$$\Delta_{c.c} = \delta k T_1. \tag{II.13}$$

In order to determine this, it is necessary to calculate a value for δ. When measuring the critical voltage we used a pulse of front length $\tau_{fr} = 10^{-3}$ sec. When the field reaches its critical value, an electron should be liberated in a time approximately equal to the observation time $\tau_{obs} = \tau_{fr}$. We have already noted that, when $E = E_{cr}$,

$$P\tau_{obs} \approx 1. \tag{II.14}$$

At the critical value of the electric field, equation (I.5) may be written in the form

$$\frac{1}{\tau_{obs}} = ae^{-\delta}. \tag{I.5'}$$

Taking logarithms, we have

$$\delta = \ln a + \ln \tau_{obs}. \tag{I.5''}$$

The preexponential factor a in equation (I.5) has not been calculated, but we can obtain it from equation (I.1), into which equation (I.5) transforms automatically [9] at low temperatures, when $T \leq eEd_{c.c}/k$. The error in doing this will evidently be small, because the preexponential factor varies little with temperature and occurs as its logarithm in equation (I.5). From equation (I.1) we have

$$a = 0.48 \cdot 10^7 \frac{E}{\sqrt{\Delta_{c.c}}}, \tag{II.15}$$

where E is the field in volts per centimeter, and $\Delta_{c.c}$ is the depth of the level in electron-volts. In our case, $E \sim 10^5$ V/cm and $\Delta \sim 1$ eV, so that $a \sim 10^{12}$ sec^{-1}. Then the value of δ, calculated from equation (I.5), is 21,

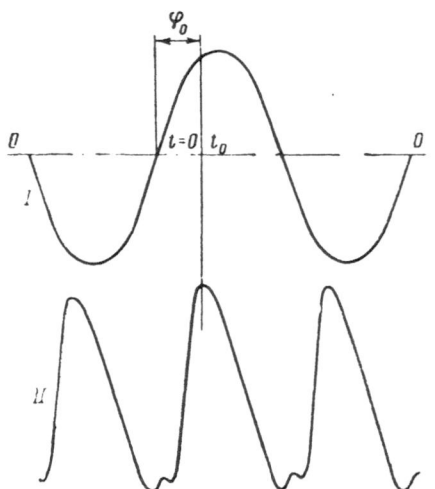

Fig. 22. Oscillograms of (I) the exciting voltage and (II) the luminance: t_0 is the instant of time and φ_0 is the phase angle corresponding to the main peak maximum of the luminance wave. Time and angle are reckoned from the instant when the voltage has zero value.

and the depth of the level, calculated from equation (II.13), is 0.710 ± 0.005 eV. If a differs from the value stated by an order of magnitude in either direction, then the depth of the level lies between the limits

$$\Delta_{c.c} = 0.635 \text{ to } 0.795 \text{ eV.}$$

According to our hypothesis, the deepest traps or donors are destroyed at the critical voltage. It is, therefore, interesting to compare the depth of level obtained above with the depth of the donor levels for the electroluminophor ZnS-Cu. The dark conductivity of an electroluminescent ZnS monocrystal has been measured [100]. With a field $E = 2 \times 10^3$ V/cm, the current density was $j = 10^{-4}$ A/cm^2. It is known that the equilibrium concentration of free carriers, in our case electrons, is given by

$$N_0 = \frac{j}{\mu e E},\qquad (\text{II.16})$$

where μ is the mobility of the electrons. Since $\mu = 120$ cm^2/ V·sec [28], it follows that $N_0 \approx 10^{12}$ cm^{-3}. This value for the equilibrium concentration of electrons in the conductivity band, with a width of the forbidden band equal to 3.7 eV, can only be produced by donors. If a small proportion of the donors is ionized, then the relation between the equilibrium concentration of electrons and the depth of the donor levels has the form

$$N_0 = \sqrt{N_d \, N_{eff}} \; e^{-\frac{\Delta_d}{2kT}},\qquad (\text{II.17})$$

where N_d is the donor concentration; Δ_d is the depth of the donor level; $N_{eff} \approx 2.5 \times 10^{19}$ cm^{-3} [101, 102]. In electroluminescent monocrystals, the concentration of the Cu activator is about 10^{-5} g-atom/g-mole, i.e., $N_d \approx 10^{17}$ donors/cm^3. In this case, the depth of the donor level, determined from equation (II.17), is 0.725 eV. If the concentration of donors differs from the value given above by an order of magnitude in either direction, then the depth of the donor level will be between the limits $\Delta_d = 0.685$ to 0.810 eV. Thus the calculated depths of the donors in a ZnS-Cu electroluminophor, and of the deepest centers participating as electron sources in production of the luminance wave, are approximately equal to each other.

§ 6. Relation between the Main Peak Maximum of the Luminance Wave and of the Frequency and Amplitude of the Exciting Voltage [103]

It was shown, in the previous section, that the variation with temperature of the main peak maximum of the luminance wave conformed with our interpretation of the processes leading to its formation. We will now see whether there is anything contrary to this interpretation in the variation of the main peak maximum with the parameters of the exciting voltage, namely amplitude and frequency.

We have already noted that the luminance wave is produced by return of electrons to ionized radiation centers. The main peak of the luminance wave is produced by return of electrons located at the opposite edge of the crystal and liberated from capture centers as the result of processes controlled by the electric field. With increasing electric field there is an increase in the probability of electron liberation, and this means that there is a greater supply of electrons in the region where the ionized centers are concentrated, so that there is an increase in luminance. On the other hand, depletion of these capture centers limits the increase in luminance at some particular voltage, i.e., the luminance wave passes through a maximum. It has been stated above that at

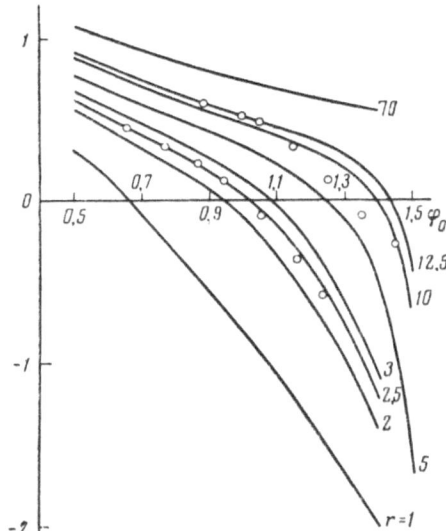

Fig. 23. Relation between the logarithm of the amplitude of the electric voltage and the phase angle φ_0 corresponding to the main peak maximum of the luminance wave. Plotted as ordinates are $\log \chi$ for the theoretical curve and $\log U_0$ for the experimental curve. The continuous curves were calculated from equation (II.23) and correspond to different values of the parameter $r = r_0 \omega$, while the circles represent experimental points.

$T \gtrless 280°K$ this liberation of electrons occurs mainly as the result of tunneling transitions, with ionization of the capture centers by the simultaneous action of heat and field. This phenomenon is analogous to thermal luminescence, with the only difference that the probability of liberation is determined by the simultaneous action of heat and field, instead of by heat alone. Thus we can calculate a theoretical relation for the phase of the main peak maximum of the brightness wave and compare it with the experimental relation for this temperature region. *

Analysis of the luminance wave shows that the maximum of the main peak occurs at a luminescence less than half of the light sum (Fig. 22). The probability of electron liberation changes by several orders of magnitude from the instant of switching on the voltage ($t = 0$) to the instant of this maximum (t_0), since it varies rapidly with the electric field. In view of what has been said above, we will, in the calculations, neglect the change in number of ionized radiation centers in the time interval $0 - t_0$.

Thus we can argue that the luminance $B(t)$ at the instant of time t is proportional to the decrease in electrons in the capture centers. This decrease is equal to the probability $P(t)$ of liberation by tunneling at this moment multiplied by the concentration of localized electrons $N_L(t)$, i.e.,

$$B(t) \sim \left[- \frac{dN_L(t)}{dt} \right] = P(t) N_L(t). \qquad \text{(II.18)}$$

Integrating, we obtain

$$N_L(t) = N_L^0 e^{-\int_0^t P(\tau)\,d\tau}, \qquad \text{(II.19)}$$

where $N_L^0 = N_L(0)$.

Substituting (II.19) in (II.18), we obtain

$$B(t) \sim N_L^0 e^{-\int_0^t P(\tau)\,d\tau + \ln P(t)}. \qquad \text{(II.20)}$$

The maximum luminance corresponds to the time for which the exponent is a maximum. The conditions for maximum luminance is therefore obtained by equating the differential of the exponent to zero:

$$P^2(t_0) = P'(t_0) \qquad \text{(II.21)}$$

where P' denotes the derivative of P with respect to time.

---

* Such a comparison is more difficult in the low-temperature region, where $T \lessgtr 220°K$ and where there is evidently considerable liberation of electrons by impact ionization. Cascade processes can then develop, so that the number of electrons liberated will not only depend on the probability of impact ionization, but also on the shape and dimensions of the region in which the electric field is concentrated.

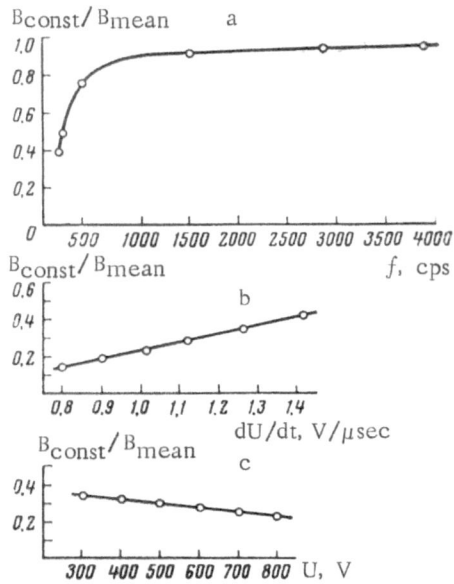

Fig. 24. Variation of the ratio of the constant luminance component to the mean luminance of electroluminescence (a) with the frequency, (b) with the slope of the front, and (c) with the pulse amplitude of the trapezoidal exciting voltage.

Now we can substitute the W, the probability of electron liberation by tunneling under the simultaneous action of the field and thermal vibrations of the crystal lattice, from equation (I.5) and the expression for the coefficient (II.15). In making this substitution we assume, first, that the field at the edge part of the crystal is proportional to the square root of the applied voltage, i.e., that $E = \varkappa_1 \sqrt{U}$, where $\varkappa_1$ is a proportionality coefficient, and second that we shall use a sinusoidal voltage, i.e., that $U(t) = U_0 \sin \omega t$, where ω is the angular frequency. The result of this substitution is to give a transcendental equation relating the phase angle $\varphi_0 = \omega t_0$, * corresponding to the maximum of the luminance wave, with the terms r and χ, one of which is proportional to the frequency and the other to the amplitude, i.e.,

$$r = r_0 \omega; \quad \chi = \chi_0 U_0. \tag{II.22}$$

The transcendental equation has the form

$$\sqrt{\chi} \sin^{3/2} \varphi_0 \exp(\chi \sin \varphi_0) = r \cos \varphi_0 \left(\frac{1}{2} + \chi \sin \varphi_0 \right). \tag{II.23}$$

In this we have used the notation

$$\chi_0 = \frac{\varkappa_1^2}{24} \frac{(e\hbar)^2}{m(kT)^3}; \quad r_0 = \frac{\sqrt{\chi_{cr}} \, e^{\frac{\Delta_{c.c}}{kT}}}{a}, \tag{II.24}$$

where χ_{cr} is the value of χ corresponding† to $U_0 = U_{cr}$; a is the value of the preexponential factor [see equation (II.15)] with the critical‡ field E_{cr}.

Figure 23 shows a family of curves for various values of r, obtained by graphical solution of equation (II.23). In order to eliminate the effect of scale factor on the shape of the curves, $\log \chi$ is plotted as ordinate for the theoretical curves and $\log U_0$ is plotted as ordinate for the experimental curves. Our experiments were carried out at T = 345 to 355°K. Experimental points were obtained for two frequencies, 265 and 1325 cps. The angles φ_0 were measured directly, since, in comparison of the experimental points with the theoretical curves, it was necessary to shift the former along the ordinate axis until they coincided with one of the theoretical curves; the experimental points corresponding to the other frequency should then, without further movement, coincide with a curve corresponding to a value of r as much greater than the first as the second frequency is greater than the first. In Fig. 23, the experimental points corresponding to a frequency of 265 cps were placed on the curve for r = 2.5. The experimental points for a frequency of 1325 cps should then coincide with the curve for r = 12.5. There was, in fact, good agreement at the higher voltages, but at lower voltages the experimental points lay somewhat below the theoretical curves. Some departure of the experimental points from the theoretical curves could be expected at low voltages, since at such voltages the electrons liberated from shallow levels play a greater role than at high voltages.

---

* The phase angle φ_0 corresponding to the maximum of the luminance wave is in this case reckoned from the instant of switching on the voltage, i.e., from its zero value (Fig. 22).

† U_{cr} is determined, as in §4 of this chapter, as that value of a linearly increasing exciting voltage at which the main peak of the luminance wave passes through a maximum.

‡ E_{cr} is the value of the field in the region where it is concentrated which corresponds to the critical voltage on an electroluminescent condenser.

Let us carry out an appraisal of the depth of the level of those capture centers from which electrons are freed in the production of the luminance wave at high voltages, for which the observed deviations of the experimental points were inconsiderable. According to (II.21) and (II.23), the depth of this level is given by

$$\Delta_{c.c} = kT \ln \frac{ra}{\omega \sqrt{\chi_{cr}}} .$$

(II.25)

All the factors on the right-hand side of equation (II.25) are known experimentally, except for a. An estimate of the value of a was obtained in the previous section, namely about 10^{12} sec^{-1}. Substituting this in (II.25), together with r = 2.5 and χ_{cr} = 0.14, obtained experimentally with T = 350°K, and $\omega = 2\pi \times 265$, we obtain for the depth of the capture centers $\Delta_{c.c}$ = 0.64 eV.

It should be noted that the lack of precision in determining the experimental parameters would give an error only in the third decimal place.

The main uncertainty in calculating $\Delta_{c.c}$ is associated with estimating the value of a. If a differs from 10^{12} by an order of magnitude in either direction, then

$$\Delta_{c.c} = 0.64 \pm 0.07 \text{ eV.}$$

In the previous section the depth of these capture centers was deduced from the variation of the critical voltage with temperature and was found to be

$$\Delta_{c.c} = 0.71 \pm 0.07 \text{ eV.}$$

It is clear that the two values, calculated by independent methods, for the deepest local levels playing the role of an electron source in the formation of the main peak of the luminance wave agree quite well with each other and with the depth of the donor levels calculated from the equilibrium concentration of free electrons in an electroluminescent crystal of ZnS-Cu.

Thus, both the variation with temperature of the phase of the main peak of the luminance wave and the variation of its amplitude with the frequency of the exciting voltage confirm the hypothesis that the main peak of the luminance wave is produced by return to the ionized radiation centers of electrons freed from capture centers by processes controlled by the electric field. These processes are impact ionization at low temperatures, and ionization by simultaneous action of the field and of thermal vibrations of the lattice at high temperatures.

Finally, it must be pointed out that this theory does not apply to a luminophor for which the radiation is not due to recombination and where the luminescent light sum is not associated with movement of electrons in the crystal, but where the luminescence is the result of processes developing in all the radiation centers.

§7. The Constant Component of Electroluminescence

We have noted previously that in the electroluminescence excited by an alternating voltage there is a constant background luminance, denoted by K in Fig. 10.

Our investigation of this constant component [104] was carried out with an electroluminescent condenser in which the thickness of the working layer was 0.2 mm. Liquid melamine-formaldehyde resin was used as dielectric. The electroluminescence was excited by a trapezoidal voltage. Where not otherwise stated, the amplitude of the voltage was U_0 = 300 V, the frequency was f = 200 cps, and the slope of the voltage front was dU/dt = 1.42×10^6 V/sec.

Figure 24 shows the variation of the ratio of the constant luminance component to the mean luminance of electroluminescence with the frequency, with the slope of the front, and with the pulse amplitude of the trapezoidal exciting voltage. It is clear from Fig. 24a that the share of the constant component increased with the frequency. It was this increase which suggested that the constant component might be due to overlapping of the peaks of the luminance wave. However, an increase in the share of the constant component with increasing

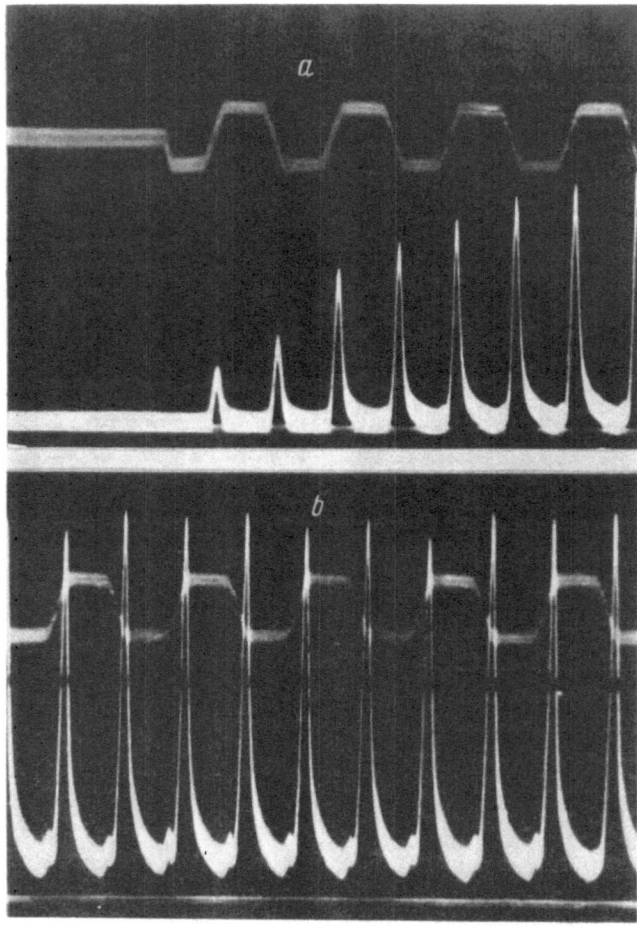

Fig. 25. Oscillograms of electroluminescent luminance:
a) from start of voltage pulses; b) 0.3 sec after switching
on. The frequency was 200 cps.

slope of the front of the exciting voltage and with decreasing amplitude of this voltage would not be expected from this point of view.

In fact, by increasing the slope of the front of the exciting voltage, or by reducing its amplitude without changing its frequency, we increased the pulse plateau and still further increased the time available to the scintillations for extinction. In other words, we created conditions for reducing the degree of overlap, but nevertheless observed an increase in the constant component.

Again, irradiation with long-wavelength light (isolated by an RG-8 filter from the spectrum of an incandescent lamp) had a pronounced effect on phosphors providing a considerable light sum; only the constant component was affected significantly, and this was reduced by about a half. With phosphors providing a small light sum, red light also reduced the alternating component, but considerably less than it affected the constant component.* There was also a marked distinction between the responses of the constant and alternating components

* In §3 of this chapter we noted the pronounced quenching action of infrared radiation on the wave luminance of electroluminescence. However, in that experiment, the intensity of the infrared radiation was greater by about three orders of magnitude.

to the time for build-up of electroluminescence. This was specially noticeable if the phosphor of the electro luminescent condenser had been well de-excited beforehand. * After the voltage had been switched on to such a condenser, the amplitude of the alternating component of the luminescence was practically fully established after 5-15 pulses, whereas the constant component became stationary after 200-400 pulses. Figure 25 shows that the constant component was hardly visible on an oscillogram at the instant when the alternating component was established.

We will now attempt to explain the results obtained. It has already been noted that excitation of electroluminescence evidently occurs in those parts of the crystal where the strongest electric field is produced. Return of electrons to these parts of the crystal leads to formation of the luminance wave. However, excitation may be extended from these to other parts of the crystal, owing to diffusion of holes and their settling down as radiation centers. Owing to polarization of the crystal, the field in the interior is less, by several orders of magnitude, than in those places where the electroluminescence is excited. Thus the effect of the voltage phase on the luminescence from the interior part should be very much less. Moreover, the concentration of ionized centers should be very much less in the interior than in the places where excitation occurs. We may therefore suppose that recombination of electrons with ionized radiation centers in the interior of the crystal gives rise to a radiation similar to photoluminescence. The existence of such a luminescence, engaging the whole volume of a crystal, was shown by experiments comparing the quenching of photo- and electroluminescence [105]. These experiments established that the whole volume of a crystal participated in the last stage of extinction, after electroexcitation and luminescence, whereas at the start the main light beam was emitted by a small part of the crystal.

We suggest that the main part of the constant component at low frequencies of the exciting voltage is associated with radiation from the interior of the crystal. If this is so, it should be possible to correlate the special features of the constant component as described above, and to give a qualitative explanation from a single point of view.

In fact, if our suggestion is correct, the constant component should take a relatively long time to become established as compared with the alternating component, since it will take some time for excitation to be transferred to the interior of the crystal.

In regions where the field is weak, there will be no liberation of electrons and holes under the influence of processes controlled by the field. Thus the action of infrared light should be more noticeable than in regions where the field is strong, where such processes are occurring intensively. Again, it is known that the action of red and infrared light is more pronounced the lower the concentration of ionized centers in the phosphor [106]. The greater influence of red light on the constant component suggests that the latter is produced in regions where, first, the field is weak and, second, there is considerably less excitation density. On the other hand, its contribution to the mean brightness is comparable with that of the alternating component; this means that it must be produced from a considerably greater volume than that giving rise to the alternating component.

In the interior of a crystal, an electric field exists practically only while the external voltage is changing, since it is during this time that a polarized charge is formed and that charge carriers move through the thickness of the crystal. The time for polarization is about 10^{-5} sec [107]. The length of a pulse plateau is about 10^{-3} sec, so that almost immediately after the external voltage stops changing, it ceases to act on the internal regions of the crystal.

During the very short time when the internal parts of the crystal are subject to a field (even if this is insufficient to liberate electrons and holes), the field will draw back into the excitation region those holes which have been liberated from radiation centers by heat. This will lead to a reduction in the volume radiation. Consequently, the longer the time for the voltage to be established, i.e., the longer the front of the pulse, the less should be the share of the constant component in the total luminescence. This is what was found experimentally (Fig. 24b, c).

---

* De-excitation was carried out either with long-wavelength light (isolated by an RG-8 filter from the spectrum of an incandescent lamp) or by preparing the electroluminescent condenser from a phosphor which had been heated in the dark.

With an increase in the pulse frequency there must be an increase in the slope of the front. This should evidently produce an increase in the share of the constant component with rise in frequency. In addition, a rise in frequency should clearly result in pulse overlap, as discussed above.

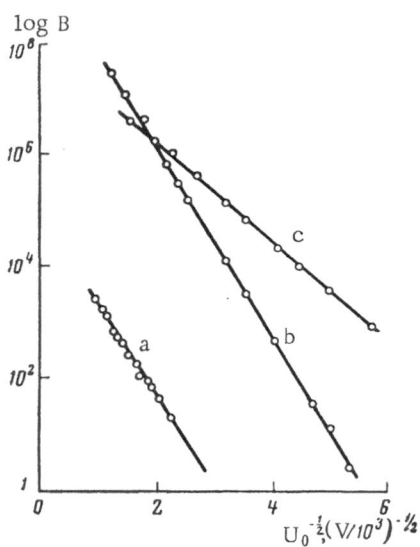

Fig. 26. Relation between the logarithm of the luminance and $U_0^{-\frac{1}{2}}$ for various electroluminophors: a) ZnS, Li_2S-Mn; b) ZnS-Cu, Mn; c) ZnS-Cu, Cl.

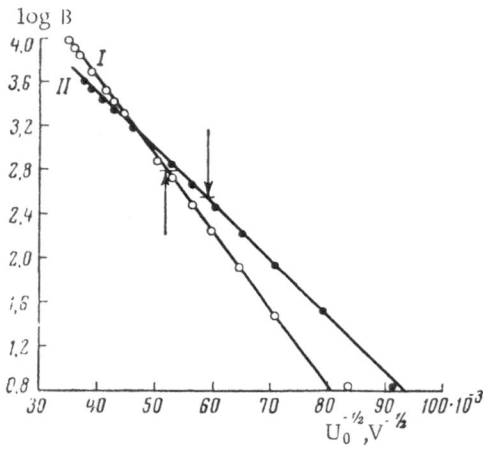

Fig. 27. Relation between the logarithm of the mean luminance of electroluminescence and $U_0^{-\frac{1}{2}}$, where U_0 is the voltage amplitude, at (I) 77°K and (II) 373°K. The arrows show the critical voltages for these temperatures.

§8. The Mean Luminance of Electroluminescence

The previous sections were concerned with the instantaneous luminance of electroluminescence. We investigated the change in luminance over a period of the exciting voltage and decided the question of when and where the radiation of the light sum created during a period in the electroluminescent condenser occurred. We will now consider the mean luminance of the electroluminescence which characterizes the value of this light sum.

The lower limit of voltage used by investigators for exciting electroluminescence depends on the sensitivity of the equipment and the thickness of the electroluminescent layer. A weak electroluminescence, with a luminance of about 10^{-10} ft-lambert, has been recorded [108] from a sublimed film about 1 μ thick with a voltage of 1.55 V. This voltage was insufficient for communicating to an electron the energy corresponding to a quantum of the light emitted. Electroluminescence at such low voltages has been explained [109] as due to the simultaneous action of heat and the field. In fact, electroluminescence, like any other form of luminescence, is the excess over the thermal radiation of a body. The application of even a small voltage to a crystal should lead to an increase in its radiation owing to the simultaneous action of heat and field (somewhat analogous to an anti-Stokes photoluminescence). Thus, generally speaking, the concept of the "threshold voltage" for electroluminescence has no meaning. The only question is the value of the luminescence.

It is normal to use a voltage which is greater than 1.55 V by a factor of tens or hundreds, and the radiation is then greater by 10 to 20 orders of magnitude and is mainly the result of the field.

The relation between the mean luminance of an electroluminescent condenser and the voltage is expressed well by the equation [28, 110]

$$B = B_0 e^{-\frac{b}{\sqrt{U_0}}}, \qquad (II.26)$$

where B is the luminance; U_0 is the amplitude of the voltage; B_0 and b are values independent of the voltage (Fig. 26). This relation is also valid for monocrystals [111].

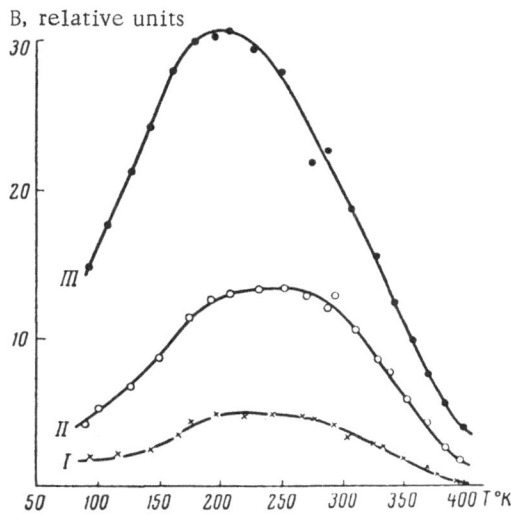

B, relative units

Fig. 28. Relation between luminance and temperature: I) for U = 140 V; II) 210 V; III) 400 V. The luminance scale is the same for curves I and II, but is five times greater for scale III.

Relation (II.26) well-represented the behavior of our electroluminophor, with both sinusoidal and trapezoidal exciting voltages, at all the temepratures investigated (77 to 413°K) and with voltages corresponding to a mean field in the electroluminescent condenser of 4×10^4 to 3×10^5 V/cm [112]. The voltages used were both below and above the critical voltages of the luminance wave. There was no change observed in the character of the relation as the voltage passed through the critical value. This is clear from Fig. 27, which shows the variation of log B with $U_0^{-\frac{1}{2}}$ for temperatures of 77 and 373°K. The critical voltages at these temperatures are marked by arrows. It is obvious that the experimental points lie on the same straight line both to the right and left of the arrows. This fact suggests that the factor determining the variation of mean electroluminescent luminance with voltage remains the same. Hence it follows that different processes determine the variation of mean luminance with voltage and the form of the luminance wave.

At the edge of a crystal, where the voltage U is concentrated as the result of formation of a Mott-Schottky type "depletion" barrier, the field $E \sim \sqrt{U}$. This relation between field and voltage is associated with the broadening of this barrier with increasing voltage, $L \sim \sqrt{U}$ [see (I.12)]. Using this relation between field and voltage, we may rewrite equation (II.26) in the form

$$ B = B_0 e^{-b_1'E} . \tag{II.26'}$$

When the width of the barrier becomes equal to the width of the crystal, it cannot increase any more, and the field in the crystal should begin to change in proportion to the voltage. This change should produce an alteration in the relation between luminance and voltage, and there should be a change from (II.26) to the new equation

$$ B = B_0' e^{-b_2'U} . \tag{II.27}$$

A change in the relation of this sort was observed with a condenser consisting of the luminophor ZnS-Cu (Cu $\approx 10^{-4}$ g-atom/g-mole) at a voltage corresponding to a mean field in the condenser of 1.5×10^5 V/cm [113]. A similar change was observed by another investigator [114], but the result was interpreted differently. The width of the "depletion" barrier, proportional to the square root of the voltage, is also dependent on the concentration of ionized donors, so that for different electroluminophors it may vary over a range of an order of magnitude. This variation evidently explains why for many electroluminophors the relation between luminance and voltage retains the form of equation (II.26) up to voltages considerably higher than at which the transition to (II.27) was observed in the first paper quoted [113].

The mean luminance of electroluminescence is strongly dependent on temperature. Figure 28 shows the

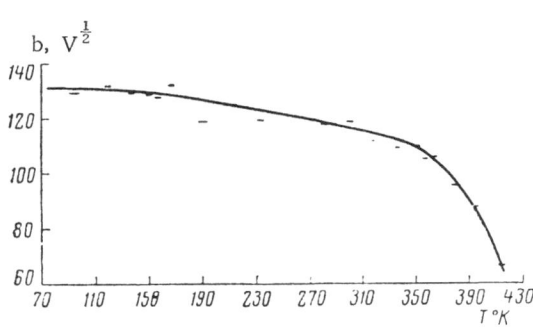

b, $V^{\frac{1}{2}}$

Fig. 29. Variation of the characteristic voltage with temperature.

results of our measurements [115]. It is clear from this figure that the mean luminance first increased with temperature, reached a maximum value, and then decreased. The characteristic voltage in equation (II.26) varied with temperature in a simpler way (Fig. 29) [112], in that it continued to decrease with rising temperature. Thus it is evident that the change in b with temperature was only partially responsible for the change in mean luminance. In fact, the change in luminance associated with the change in characteristic field should lead to a continuous increase in luminance with temperature.

More complex variations of the mean luminance and characteristic voltage with temperature have been described (e.g., [28]). We shall discuss these later.

The mean luminance of electroluminescence increases approximately linearly with the frequency of the exciting voltage up to 5-10 kcps. The increase slows up at high frequencies; the green band shows this effect at lower frequencies than the blue band does. This has been interpreted [116] as due to a decrease in the probability of transition of holes from "blue" to "green" centers with a decrease in the period of the exciting voltage.

§9. The Mechanism of the Excitation of Electroluminescence

Several investigators have come to the following conclusion [28, 39, 117, 118]:

1. Electroluminescence in ZnS is excited by impact ionization of the crystalline lattice and of luminescence centers. Direct excitation of electroluminescence by the field is impossible, since this requires a field of the order 10^7 V/cm, whereas cascade breakdown of ZnS occurs at a field of about 10^6 V/cm.

2. The relation between mean luminance and voltage given by equation (II.26) agrees with the impact ionization mechanism, since the mean luminance depends on the field (II.26') in the same way as the probability of impact ionization in the Seitz equation (I.10).

We consider that the second statement is incorrect. The Seitz equation is only applicable to ionic crystals when the field is small, i.e., $E \ll E_1^*$ [see (I.11)] [22], and the correct equation to use with fields greater than E_1^* is (I.9), which involves the second instead of the first power of the field. For the temperature range of our investigations $1.5 \times 10^4 < E_1^* \lesssim 7 \times 10^4$ V/cm, so that the free path length of an electron in ZnS is about 10^{-6} cm [28]. Even the mean field in a condenser when used to excite electroluminescence normally exceeds this value. Moreover, there is no justification for applying ionic crystal equations to impact ionization in ZnS. We have already noted that the electron affinity data correspond to a bond which is about 30% ionic in ZnS. A similar value has been obtained by comparison of optical phonons, longitudinal and transverse [119]. In addition, the free path length of an electron in ZnS is of the same order as in germanium and silicon, so it is clear that, as regards conditions for impact ionization, ZnS should be considered as more like a covalent semiconductor than an ionic one. Now with covalent semiconductors, the probability of impact ionization is given by equation (I.9). Yet the probability of impact ionization so given does not agree with the observed relation between mean luminance of electroluminescence and voltage.

On the other hand, the relation between probability of direct field ionization and the value of the field given by equations (I.1) and (I.3) is in accordance with equation (II.26') for the luminance of electroluminescence, since the field appears in the same form in both.

Let us now compare the effects of temperature on the relation between luminance and voltage and on the probabilities of field ionization and impact ionization. In the previous section we have seen that the characteristic voltage b for electroluminescence, and therefore the characteristic field b_1, decreases with rising temperature (Fig. 29). The characteristic field for field ionization by electron tunneling through a potential barrier also decreases with rise in temperature, owing to a reduction in the barrier height. On the other hand, the characteristic field for impact ionization increases with rising temperature in accordance with (I.8).

Thus, the effect of temperature on the relation between mean luminance of electroluminescence and voltage is consistent with the process of field ionization by tunneling of electrons, and is not consistent with the process of impact ionization.

Let us consider at what levels the tunneling of electrons responsible for the relation (II.26) occurs. This cannot be at the level of capture centers $\Delta_{c.c}$ in ZnS-Cu because, as shown in §5 of this chapter, the liberation

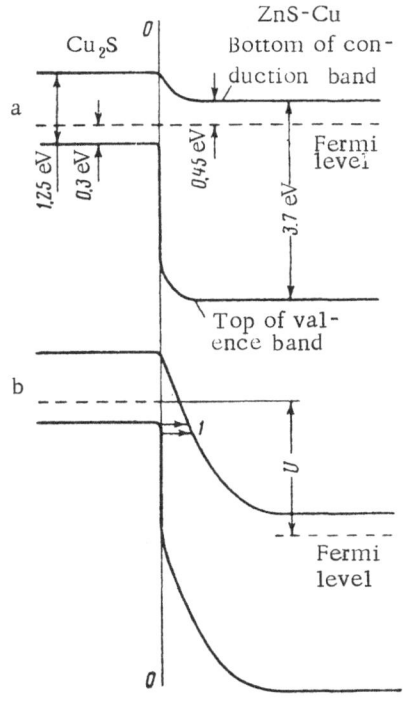

Fig. 30. Band diagram of a ZnS-Cu luminophor in contact with a Cu_2S surface phase: a) without external voltage; b) with external voltage U.

of electrons from capture centers at low temperatures takes place mainly by impact ionization. Also, these levels should have a wider energy spectrum than the levels of the capture centers of a ZnS-Cu crystal, or else a saturation of the luminance would be observed when the field reached a critical value, owing to complete destruction of these levels, similar to that observed for the wave luminance. It must be assumed that it is a tunneling transition, resulting from direct ionization by the electric field of the crystalline lattice and radiation centers. In our opinion there is no basis for the view that such ionization is unimportant as compared with impact ionization. The luminophor crystallites are about 10^{-3} cm thick, and the regions where the field is strong will be somewhat thinner than this; in this thin film, conditions can be realized such that field ionization predominates over impact ionization. However, it may also be supposed that the tunneling transitions responsible for the relation (II.26) are associated with the liberation of electrons either from surface levels having a wide spectrum or from a surface phase, such as copper sulfide, deposited on the surface of ZnS-Cu crystals.

We noted in §1 of this chapter that it has been stated in the literature that the existence or absence of electroluminescence from luminophors based on zinc sulfide is associated with the presence or absence of a surface phase. On the other hand, it has been shown [120] that it is the accumulation of electrons participating in the process which determines the increase in the mean luminance of electroluminescence after the switching on of the voltage.

Figure 30 shows an approximate band diagram of a ZnS-Cu luminophor with a Cu_2S surface phase. This is explained more fully in Appendix II. The arrows 1 indicate the penetration of electrons from the valence band of Cu_2S to the conduction band of ZnS.

It is clear from the diagram that it is necessary to establish a potential difference U ⪖ 0.75 V for tunneling transitions to proceed. It should be noted that tunneling transitions of this type occur in tunnel diodes through a boundary region with electron and hole conductivity. Equation (I.1) then applies. The calculations in Appendix III show that this equation is applicable to our case.

It is natural to suppose that the quasi-momentum of an electron is not the same at the top of the Cu_2S and ZnS valence bands and at the bottom of the ZnS conduction band, so that the transition is indirect.*

We have already noted in Chapter I that indirect transition involves participation of phonons of definite magnitude and depends on the number of such phonons. Thus equation (I.3) differs from (I.1) in having the factor

$$N_{\omega_q}(T) = \frac{1}{e^{\frac{\hbar\omega_q}{kT}} - 1}.$$ (II.28)

The variation of (I.3) with temperature depends not only on the change in the characteristic field, but also on the change in this factor.

_____

* The structure of the ZnS band is at present unknown, so that we cannot wholly exclude the possibility of a direct transition. However, we have only considered the case of an indirect transition, which is more general.

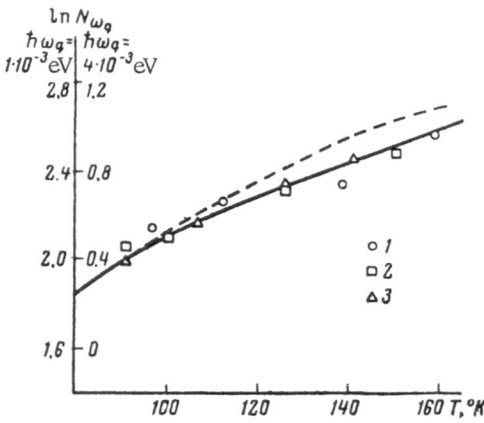

Fig. 31. Comparison of the variation with temperature of the preexponential factor B_0 and of $N_{\omega_q} = \dfrac{1}{e^{\frac{\hbar \omega_q}{kT}} - 1}$. The experimental points are for 1) U = 140 V; 2) U = 210 V; 3) U = 400 V. The continuous curve of $N_{\omega_q}(T)$ is for $\hbar \omega_q = 1 \times 10^{-3}$ eV, and the dotted curve for $\hbar \omega_q = 4 \times 10^{-3}$ eV.

We compared the variation with temperature of the luminance of electroluminescence (Fig. 28) with the variation with temperature of the probability of field ionization by indirect tunneling (I.3) [115]. This comparison was made only for the low-temperature branch of Fig. 28; comparison for the high-temperature branch is not yet possible. The decrease in luminance at higher temperatures is evidently the result of a process similar to the quenching of photoluminescence, and it is probable that holes are liberated more intensely with rising temperature. However, the variation with temperature of the probability of their liberation in electroluminescence is unknown. This process is evidently greatly diminished in the low-temperature region, and it is possible to carry out the above-mentioned comparison.

The variation with temperature of the characteristic voltage for electroluminescence has already been discussed (Fig. 29). We will, therefore, separate the preexponential factor in the luminance equation (II.26) as given by

$$B_0(T) = B(T) e^{\frac{b}{\sqrt{U}}} \tag{II.29}$$

and compare its variation with temperature with that of the preexponential factor N_{ω_q} (II.28).

Figure 31 shows the result of this comparison. The comparison was carried out with semilogarithmic paper to avoid the effect of the scale factor. Theoretical curves were calculated for values of $\hbar \omega_q$ equal to 1×10^{-3} and 4×10^{-3} eV. Experimental results for the various voltages were superposed on the theoretical curves by shifting the former, in the ordinate direction only, by the same amount. Agreement between the experimental points and the theoretical curves is worse for $\hbar \omega_q = 4 \times 10^{-3}$ eV than for $\hbar \omega_q = 1 \times 10^{-3}$ eV. Thus the value of the phonon participating in the process is not greater than 4×10^{-3} eV. The difference between the curves for $\ln N_{\omega_q}$ at low values of $\hbar \omega_q$ is less than the scatter of the experimental points, so it is not possible to determine a lower limit for the value of this phonon.

Thus, the variation with temperature of the relation between mean luminance of electroluminescence and voltage corresponds to the process of electron tunneling when the crystal is placed in an electric field. In other words, the mean luminance is determined by the number of electrons participating in the electroluminescence, and not by the conditions of their acceleration.

Finally, this tunneling is not the only process occurring in electroluminescence. We have already stated that a whole series of processes takes place during electroluminescence: concentration of the electric field; ionization of the crystal lattice itself and of luminescence centers, both directly and by impact ionization; impact excitation of luminescence centers; liberation of electrons and holes from their trapping centers [116]; transfer by the field of electrons and holes through the crystal, and also between the various phases of the crystal, if they exist. The problem is to distinguish the relative effects of these processes on the relation between luminance of electroluminescence and voltage.

The diversity of these processes evidently accounts for some of the discrepancies, noted in the preceding section for certain luminophors, between the variation with temperature of the relation between luminance and characteristic voltage and the variations shown in Figs. 28 and 29.

It is indeed possible to achieve conditions such that the mean luminance will be determined, not by the number of electrons participating in the process, but by the conditions of their acceleration. It is only necessary to note that in this case the relation between mean luminance and voltage will not be given by equation (II.26).

§ 10. Absorption of Energy during Electroluminescence

In the previous sections we have considered investigations on the mechanism of the electroluminescence of an electroluminescent condenser in terms of luminescence, i.e., the energy radiated by it. Investigation of the energy absorbed also provides a means for studying the kinetics of the process. It is therefore necessary to compare the results obtained by both methods.

When measuring the mean power absorbed by an electroluminescent condenser, some widely used methods depend on measurement of the loss tangent by some means or other (e.g., [54, 121]). We will call these bridge methods. Such methods are precise if the condenser under investigation is a linear element. However, the process of electroluminescence is essentially nonlinear. Strictly speaking, the distributions of electrons in a crystal by energy levels are quite different during electroluminescence and in the absence of a field. The conductivity therefore depends very much on the value of the field and the time for which it is applied, so that electroluminescent condensers behave as nonlinear elements [107]. Figure 32 shows an oscillogram of the active component of the current through an electroluminescent condenser excited by a sinusoidal voltage. It is clear from the figure that the current is far from sinusoidal. Hence it follows that the above methods for measuring mean power absorption are incorrect in principle, and are not at all suitable for measuring instantaneous power.

We have already said in §2 that for measuring the power consumed by an electroluminescent condenser we used the power oscillator of a loop oscillograph. The advantage of this method is the direct multiplication of

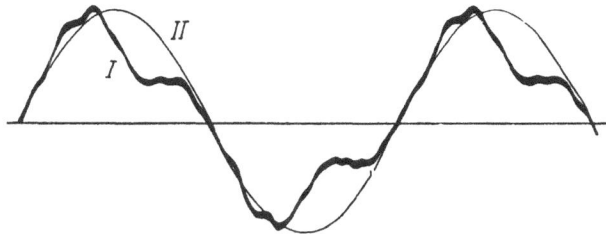

Fig. 32. Oscillograms of (I) the active component of the current through an electroluminescent condenser and (II) the exciting voltage $U_0 = 300$ V.

the instantaneous values of the current passing through the electroluminescent condenser by the applied voltage, so that it is suitable for any current wave form.

The energy absorbed in electroluminescence is consumed in moving electrons in the field created by the applied difference in potential.

In the experiments on wave luminance discussed in §§ 5 and 6 it was shown that in electroluminescence the free electrons arise mainly by liberation from capture centers. This means that for approximate calculations we can neglect the dark conductivity and the number of electrons liberated by ionization of the crystal lattice and radiation centers. The energy taken by an electron from the field is equal to the product of the charge moved and the difference in potential through which it moves. Thus, since free electrons arise mainly at the crystal edges, where the field is a maximum according to the Mott-Schottky model, they move through approximately the same difference of potential U. Thus the field energy consumed by all these electrons is approximately

$$I \approx eNU, \tag{II.30}$$

where N is the number of free electrons.

The number of these electrons will increase with the voltage in a way which will depend on how the electrons are liberated from the capture centers. In §5, on the variation of the wave luminance with temperature, it was shown that at high temperatures (T $\gtrsim$ 280°K) electrons are released from traps by tunneling with the participation of many phonons, and that the probability of release depends on the field in accordance with equation (I.5). As before, equation (II.15) can be used for the preexponential factor a. In the case of the crystal edge, where E $\sim$ U, equation (I.5) may be rewritten as

$$P_t^T = a_1^0 \sqrt{U} e^{b_1^0 U}, \tag{II.31}$$

where a_1^0 and b_1^0 have values which do not depend on the voltage.

Assuming that the number of free electrons is proportional to the probability of their release from traps, equation (II.30) can be rewritten for this case in the form

$$I_t \sim a_1^0 U^{3/2} e^{b_1^0 U}. \tag{II.32}$$

At low temperatures (T $\lesssim$ 220°K), electrons are mostly released from traps by impact ionization. The mean probability of impact ionization is related to the field by equation (I.9), where n depends on a number of factors, chief among which are (1) the variation of the free path of an electron with its energy and (2) the extent of the effect of impact ionization on the distribution function of electrons with respect to energy in the region of higher ionization potential. Both these factors are unknown for the case of ZnS. However, since the preexponential term in (I.9) varies only slightly with the field as compared with the exponent itself [18, 19], we will take it as being constant. Then for the edge part of the crystal equation (I.9) may be rewritten as

$$P_{imp} \approx a_2^0 e^{-\left(\frac{b_2^0}{U}\right)}, \tag{II.33}$$

where a_2^0 and b_2^0 have values which do not depend on the voltage. Assuming that the number of free electrons released by impact ionization is again proportional to the probability of their release, * equation (II.30) may be rewritten as

---

* In the release of electrons by impact ionization this assumption is less obvious than in the case of release by tunneling, because with the development of cascade processes the number of freed electrons will depend not only on the probability of impact ionization, but also on the shape and dimensions of the region where the voltage is concentrated.

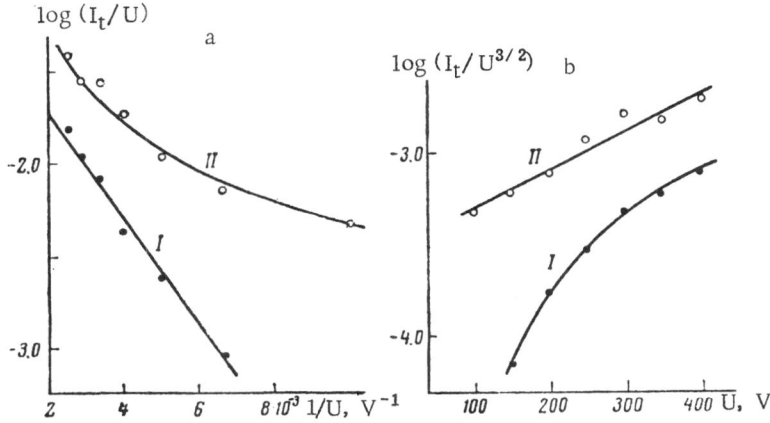

Fig. 33. Dependence of the energy absorbed by an electroluminescent condenser on the voltage: I) for T = 114°K; II) for 399°K.

$$I_t \sim a_2^0 U e^{-\left(\frac{v_2^0}{U}\right)} . \qquad (II.34)$$

Equation (II.34) differs considerably from (II.32). Thus, having measured the relation between I_t and U at different temperatures, we can check the concepts developed in our investigation of the wave luminance.

To facilitate the comparison of equations (II.32) and (II.34) with experiment, we take logarithms and re-write the equations as follows:

$$\log\left(\frac{I_t}{U^{3/2}}\right) = \log a_1^0 + \frac{b_1^0}{2.3}U, \qquad (II.32')$$

$$\log\left(\frac{I_t}{U}\right) = \log a_2^0 - \frac{b_2^0}{2.3}\frac{1}{U} . \qquad (II.34')$$

Figure 33 shows the results of the measurements. It is clear that $\log(I_t/U)$ is a linear function of U^{-1} at lower temperatures (Fig. 33a), and that $\log(I_t/U^{3/2})$ is a linear function of U at higher temperatures (Fig. 33b).

Thus, the variation of the energy consumed with voltage is consistent with the postulates that at low temperatures electrons are released from traps mainly by impact ionization, and at high temperatures electrons are released by simultaneous action of the field and heat.

§11. The Energy Yield of Electroluminescence

Measurement of the absolute energy yield as a function of various factors is one of the most important methods for investigating the processes within a luminescent body. Such measurements are also clearly of practical importance for investigations on the possibility of using luminescent light sources.

Electroluminophors synthesized up till now give relatively low energy yields. The best light yield of a ZnS-Cu electroluminophor described in the literature is 4% [54]. This is much the same as the light yield from a modern incandescent lamp, whereas the efficiency of a luminescent lamp, as now used for illumination, is somewhat higher.

We noted in the previous section that it is normal to use a bridge method for measuring the energy absorbed in electroluminescence, although its use in this case is incorrect. It is therefore of interest to compare our results with data obtained by the bridge method.

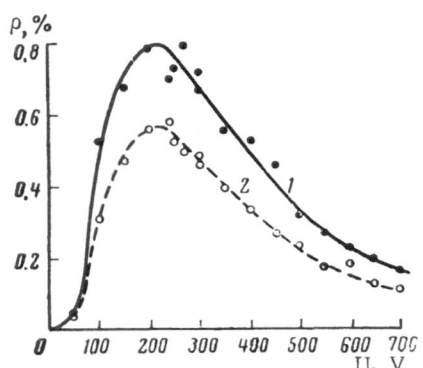

Fig. 34. Variation of the electroluminescence yield with the voltage: 1) measured with the power oscillator; 2) measured by the bridge method.

We carried out such measurements [122] on electroluminescent condensers with luminophors from three different sources: a ZnS-Cu, Al luminophor, prepared by R. M. Medvedeva at the Physics Institute of the Academy of Sciences, as used in our previous work; a ZnS-Cu luminophor kindly submitted to us by F. M. Pekerman; a No. 2 ZnS-Cu green luminophor from the Institute of Vacuum Electrotechnology in Prague. All these phosphors also contained chlorine. Solid condensers were prepared from the phosphors by the method described in §2 of this chapter. Excitation was by means of a sinusoidal voltage with a frequency of 50 cps (from the mains). All the measurements were carried out at room temperature.

The mean power absorption was measured by the oscillographic and bridge methods for each value of the voltage. The light energy radiated was measured at the same time. The luminance of the luminescence was recorded by an FÉU-19 photomultiplier, which had previously been calibrated in absolute energy units. The photomultiplier was calibrated against a thermopile, whose constant had been determined against a Hefner standard lamp. In calculating the energy radiated by the electroluminescent condenser we took account of the angular distribution, which according to our measurements was expressed by a cosine law. The precision of the relative measurements of the power radiated was 1%, but the precision of the absolute measurements was 20%. The ratio of the energy radiated I_r to the energy absorbed I_{ab} gave the value of the absolute energy yield ρ. The precision of the relative yield measurements was 6%, but the precision of the absolute measurements was 30%.

Figure 34 shows the observed relation between yield ρ and voltage U (curve 1). The results were obtained with an electroluminescent condenser made up from the luminophor from the Physics Institute of the Academy of Sciences; the results with the other luminophors were similar. It is clear from the figure that the yield at first increased sharply with voltage, passed through a maximum, and then dropped.

Curve 2 of Fig. 34 shows the relation between yield and voltage when the power absorbed was measured by a bridge method. It follows that the bridge method gave a lower value for the yield as compared with the oscillographic method, and the difference considerably exceeded the relative error of either method. This difference depended on the voltage and was up to 40%.

In order to obtain additional confirmation that this difference really was due to the nonsinusoidal character of the current through the electroluminescent condenser and was not due to any defect in calibration, we measured the absorption of energy in the same ohmic resistance by both methods and obtained identical results.

A relation between light yield and voltage similar to that obtained by us was obtained by other workers [123] using the bridge method. According to their results, the light yield at the maximum of the curve was 10 lumens/watt. The maximum energy yield which we observed was 1.3%, which corresponds, for green light, to a light yield of about 6 lumens/watt.

This relation has been measured [54] for phosphor particles of different dimensions. It was found that the maximum light yield varied approximately as the reciprocal of the square root of the particle size.

We have compared [122] our observed relation between yield and voltage with the theoretical relationship based on the concepts of energy absorption discussed in the previous section. According to these concepts, the absorption of energy at $T \gtrsim 280°K$ (which includes room temperature) should vary with the voltage in accordance with equation (II.32). The energy radiated is proportional to the luminance of electroluminescence and, consequently, should vary with the voltage in accordance with equation (II.26). In this case, the relation between yield and voltage should be given by

$$\rho(U) = \frac{a_3^0}{U^{3/2} \exp\left(b_1^0 U + \dfrac{b}{\sqrt{U}}\right)}, \qquad \text{(II.35)}$$

where the value of a_3^0 is independent of the voltage.

The characteristic voltage b was calculated from the relation between luminance and voltage. For the luminophor from the Physics Institute of the Academy of Sciences $b = 92$ $V^{\frac{1}{2}}$. The value of b_1^0 was obtained from the relation between temperature and critical voltage for the wave luminance (see §5). U_{cr} at room temperature was about 600 V, and the depth of the capture centers was 0.71 ± 0.08 eV, so that $b_1^0 = 0.01 \pm 0.004$ V^{-1}.

To eliminate the effect of the scale factor, the relation between ρ and U was plotted in terms of logarithmic coordinates. Figure 35 shows the experimental results for ρ as a function of U for all the samples. Since the maxima of the $\rho(U)$ curves differed somewhat for the different luminophors, we superposed them on the

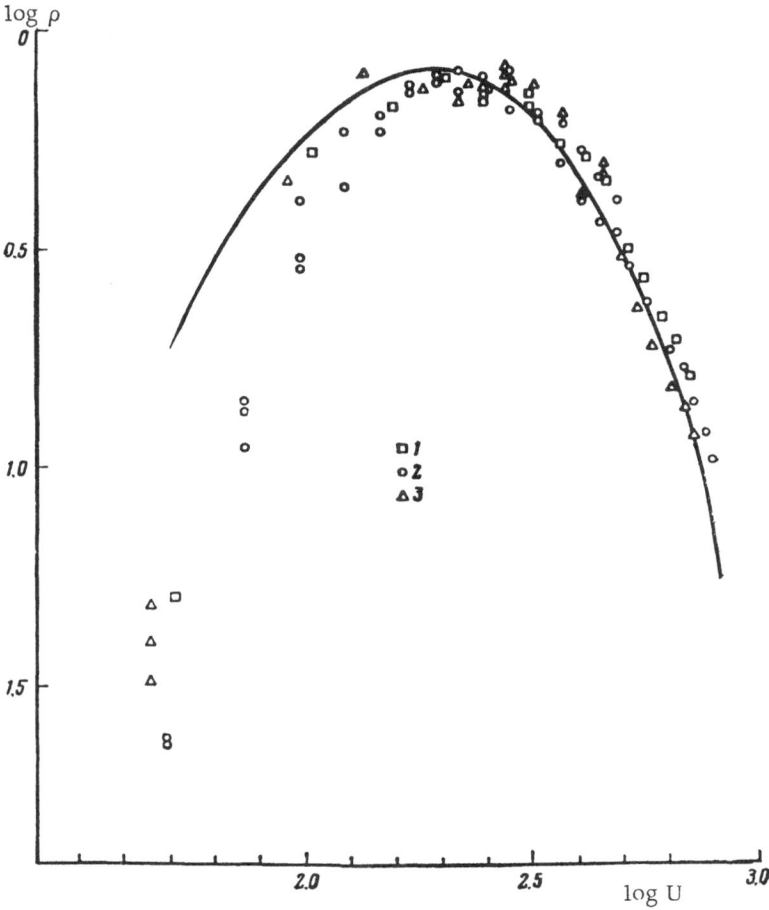

Fig. 35. Comparison of the experimental relation between yield and voltage with the theoretical relation of equation (II.35): 1) data for luminophor from the Physics Institute of the Academy of Sciences; 2) luminophor from F. M. Pekerman; 3) Czechoslovak luminophor. The continuous curve is the theoretical.

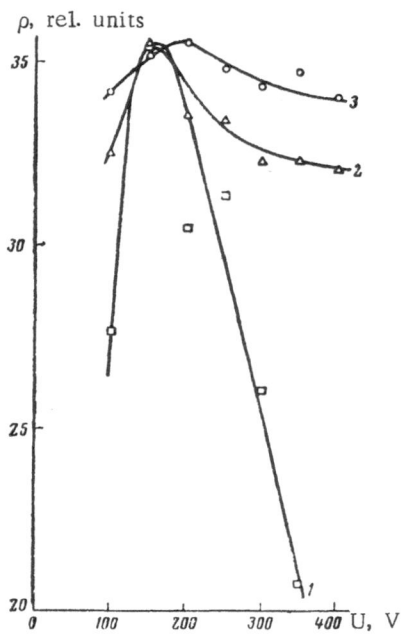

Fig. 36. Relation between electroluminescence yield and voltage at the following temperatures: 1) 294°K; 2) 114°K; 3) 399°K. The origin has been moved up 29.1 units for curve 2 and 31.6 units for curve 3.

maximum for the luminophor from the Physics Institute of the Academy of Sciences; in most cases this involved a displacement along the direction of the abcissa axis equivalent to a change of scale by a factor of 1.25. The theoretical curve was superposed on the experimental points by moving it along the direction of the ordinate axis only.

It is clear from the figure that the theoretical curve agreed well with the experimental points at high voltages, but was not steep enough at the lower voltages. This difference is evidently associated with the fact that in calculating the relation (II.35) we did not take into account the size distribution of luminophor particles. In fact, the steepness of the $\rho(U)$ curve depends very much on the particle size [54]; to be more precise, it decreases with increasing particle size.

We investigated [115] ρ as a function of U not only at room temperature, but also at 114 and 399°K, corresponding to the low- and high-temperature branches of the curve in Fig. 19. At these temperatures $\rho(U)$ also has maxima. Figure 36 shows the results of these measurements. The curves corresponding to the temperatures 114 and 399°K have been shifted in the direction of the ordinate axis so that their maxima are at the same level as the maximum of the room temperature curve. It is clear from the figure that the position of the maximum of the $\rho(U)$ curve shifts towards a higher voltage with increase in temperature.

Both 294 and 399°K are temperatures within the range for which equation (II.35) was derived. We can therefore compare the observed displacement of the maximum of the $\rho(U)$ curve on changing from one temperature to the other with the displacement calculated from this equation. Finding the maximum of the function (II.35) leads to the condition

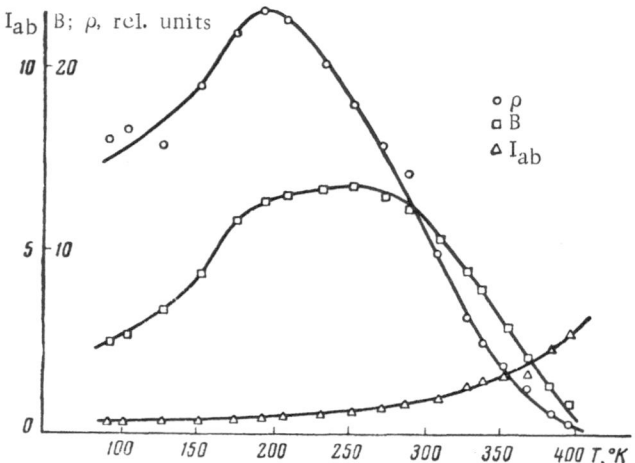

Fig. 37. Variations with temperature of the energy absorbed by an electroluminescent condenser I_{ab}, the luminance of the luminescence B, and the yield ρ.

$$\frac{3}{2} U_M'^{1/2} + b_1^0 U_M^{3/2} - \frac{b}{2} = 0. \tag{II.36}$$

With the condenser investigated, U_M for T = 294°K was 160 V (Fig. 36), and b, determined from the expression for B(U), was 80 $V^{1/2}$. The value of b_1^0, calculated for these values of U_M and b from equation (II.36), is about 0.01. For the other temperature (T = 399°K), b = 70. It is known that b_1^0 varies approximately as $1/T^3$ [see equation (I.5)], so that b_1^0 for this temperature is about 0.004. Then from the conditions (II.36) we have that, at T = 399°K, $U_M \approx 215$ V. From the experiments at this temperature U_M = 200 ± 20 V.

We also investigated the variation with temperature of the yield at various voltages. The curve for $\rho(T)$ also had a maximum (Fig. 37). The temperature T_M at the maximum of the $\rho(T)$ curve decreased with increasing voltage; the table below shows some results obtained with one of the condensers containing our luminophor:

U, V	140	210	400
T_M, °K	200	190	160

The presence of a maximum in the $\rho(T)$ curve is clearly associated with the maximum in the B(T) curve.

An investigation has been made [123] of the relation between electroluminescent yield and frequency for ZnS-Cu luminophors. It was found that the form of this relation was different for two types of luminophor which differed as regards the relation between luminance and frequency. In the case of those luminophors whose luminance increased linearly with frequency, * the yield was practically independent of frequency.

In other cases,† the yield at first increased with frequency, reached a maximum, and then decreased. The frequency corresponding to the maximum yield increased with increasing voltage.

---

* For example, such a relation between luminance and frequency is normally observed with electroluminophors up to a frequency of about 10 kcps (see §8).

† This particularly applies to electroluminophors which show strong temperature quenching.

CONCLUSION

In conclusion it should again be emphasized that an electroluminescent condenser is a rather complicated electrical system. The fact that in spite of this it is possible to obtain quite good results from calculations based on a relatively simple model can obviously be explained by the very marked dependence on voltage of those processes which form the basis of the phenomenon of electroluminescence.

There is no doubt that investigation of monocrystals would substantially improve our understanding of this phenomenon. However, it is also clear that knowledge obtained in this way could not be applied without modification to powder luminophors. This is even more true of the application of information obtained by investigations on other substances. Nevertheless, the processes considered in Chapter I play an important role in all cases.

It is my pleasant duty to thank M. V. Fok for reading the manuscript and for valuable suggestions.

APPENDIX I

Estimation of the Oscillator Strength for Transitions Corresponding to the Release of Electrons and Holes from Trapping Levels by Infrared Light

It is clear from the oscillograms of Fig. 13 that after switching on the pulse lamp the intensity of its radiation increased approximately linearly, reaching a maximum after time t^i_M. The pulse of luminescence (peak G) produced by the infrared light reached its maximum (under the conditions used) after some shorter time t, a time which evidently sufficed to exhaust the stock of trapped electrons and holes. This made it possible to estimate the oscillator strength for transitions corresponding to the release by infrared light of electrons and holes from capture levels.

Figure 38 represents a band diagram, showing all the elementary acts considered in our calculations. Part I of this figure refers to that region of the crystal where the ionized radiation centers are concentrated, and from which there is a loss of electrons. Part II of the figure refers to the crystal region where there is a gain of electrons from region I. In region I there is a positive space charge and in region II there is a negative space charge, so that a polarization field is created in the crystals and this is capable of moving free charges (electrons from region II to I, and holes in the opposite direction).

Let us use the following nomenclature:

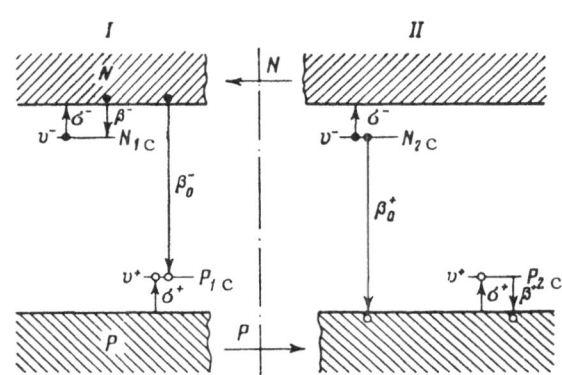

Fig. 38. Diagram of electron transitions in the electroluminescence of ZnS.

I_{ir} is the intensity of infrared light which, as already noted, before it reaches its maximum value increases approximately linearly with time, i.e., $I_{ir} \approx \alpha t$; N and P are the concentrations of free electrons and holes; υ^- is the concentration of electron capture centers, and υ^+ is the concentration of luminescence centers (hole capture centers); N_{1c} and N_{2c} are the concentrations of captured electrons in regions I and II respectively; similarly P_{1c} and P_{2c} are the concentrations of ionized luminescence centers; σ^- and σ^+ are the cross sections for release, under the influence of infrared light quanta, of electrons from capture centers and of holes from ionized luminescence centers.

$\beta^- \upsilon^-$ and $\beta^+ \upsilon^+$ are the probabilities of electron capture by capture centers and of hole capture by luminescence centers.

β_0^- is the coefficient of recombination of an electron with an ionized luminescence center (radiative recombination).

β_0^+ is the coefficient of recombination of a hole with a filled capture center (radiationless recombination).

The arrows on the diagram show the direction of electron transitions.

To simplify the solution of the problem, we consider only one kind of luminescence center and one kind of capture center. *

Repeated capture of electrons in region II and of holes in region I can obviously be neglected, since under the influence of the polarization field any free electrons will rapidly move out of region II into region I and any free holes will move from I to II.

For this reason we need not consider recombination of electrons with ionized luminescence centers in region II or of holes with filled capture centers in region I.

The kinetic equations corresponding to this scheme are

$$\alpha\sigma^- t \left(N_{1c} + N_{2c}\right) = \left(\beta_0^- P_{1c} + \beta^- v^-\right) N,$$

$$\alpha\sigma^+ t \left(P_{1c} + P_{2c}\right) = \left(\beta_0^+ N_{2c} + \beta^+ v^+\right) P,$$

$$\frac{dN_{1c}}{dt} = -\alpha\sigma^- N_{1c} t + \beta^- v^- N,$$

$$\frac{dN_{2c}}{dt} = -N_{2c} \left(\alpha t \sigma^- + \beta_0^+ P\right),$$

$$\frac{dP_{1c}}{dt} = -P_{1c} \left(\alpha t \sigma^+ + \beta_0^- N\right),$$

$$\frac{dP_{2c}}{dt} = -\alpha\sigma^+ P_{2c} t + \beta^+ v^+ P.$$

(A.I.1)

A much simpler solution can be achieved by using a completely symmetrical system for electrons and holes, i.e., by writing

$$N_{2c} = P_{1c}; \quad \beta_0^+ = \beta_0^- = \beta_0; \quad \sigma^+ = \sigma^- = \sigma; \quad \beta^+ = \beta^- = \beta;$$

$$v^+ = v^- = v; \quad N = P.$$

In this case the system of equations (A.I.1) can be rewritten as

$$N = \frac{\alpha\sigma t \left(N_{1c} + N_{2c}\right)}{\beta_0 N_{2c} + \beta v},$$

$$\frac{dN_{1c}}{dt} = -\alpha\sigma t N_{1c} + \beta v N,$$

(A.I.2)

$$\frac{dN_{2c}}{dt} = -\alpha\sigma t N_{2c} - \beta_0 N N_{2c}.$$

The maximum luminescence will be achieved at the instant t_M, which can be found from the equation

$$\frac{d}{dt} \left(N N_{2c}\right) = 0.$$

(A.I.3)

By simple calculation it can be shown from equations (A.I.2) and (A.I.3) that

$$t_M^2 = \frac{1}{2\varkappa} \frac{1}{2 + y},$$

(A.I.4)

---

* There is a sound basis for this in the case of luminescence centers, since the luminescence is mainly green.

50

where

$$x = \frac{\alpha \sigma}{2}; \quad y = \frac{\dfrac{\beta_0 (N_{1C})_M}{3v} - 1}{\dfrac{\beta_0 (N_{2C})_M}{3v} + 1}.$$ (A.I.5)

and $(N_{1C})_M$ and $(N_{2C})_M$ are the values of N_{1C} and N_{2C} at the instant of time t_M.

Using the fact that $N_{1C} \leq N_{2C}$, it can be shown that the value of y is within the limits

$$-1 \leqslant y \leqslant \frac{1}{8},$$

and accordingly that the value of t_M is within the limits

$$\frac{0.48}{\sqrt{x}} \leqslant t_M \leqslant \frac{0.71}{\sqrt{x}}.$$ (A.I.6)

The value of t_M can be obtained from experiment. We can therefore make use of the relation (A.I.6) to estimate the cross section σ for release of electrons and holes from their corresponding capture levels under the influence of infrared light quanta. From (A.I.6) the mean value of t_M is $0.6/\sqrt{x}$. Inserting this in (A.I.5), we have

$$\bar{\sigma} \approx \frac{0.7}{\alpha t_M^2}.$$ (A.I.7)

Our experimental value for t_M was approximately 0.2 msec. The energy incident on each square centimeter of electroluminescent condenser during a period of irradiation by infrared light is about 1 joule. Taking into account the intensity of the infrared light and the geometrical parameters of the oscillogram, we obtain for α a value of approximately 4×10^{25} quanta/cm$^2 \cdot$ sec^2. Then $\bar{\sigma}$ is of the order of magnitude 10^{-18} cm^2/quantum. From this cross section we can easily obtain the oscillator strength from the equation

$$\bar{F} \, dW = \frac{137 m^* \sigma dW}{2\pi^2 \hbar^2 g},$$ (A.I.8)

where $\bar{F}$ is normalized to unit energy interval of the oscillator strength; g is the statistical weight of the state to which the electronic transition occurs.

It is known that zinc sulfide phosphors are self-quenching and have, when in the excited state, additional absorption over two wide spectral regions with maxima at about 800 and 1300 mμ [124, 125].

The band in the 1300 mμ region lies outside the range over which luminescent radiation is emitted (Fig. 12). The main absorption, therefore, is at the band with a maximum at 800 mμ, extending approximately from 700 to 1000 mμ, which corresponds to $dW \approx 0.55$ eV. The oscillator strength, calculated from equation (A.I.8),* is given by $\bar{F} dW \approx 5 \times 10^{-3}$, i.e., it is relatively small. The obvious explanation is that, under the influence of infrared light quanta with an energy of about 1.5 eV, electrons and holes pass over into a state with a high momentum.

---

* In this calculation we have assumed m^* is equal to the mass of an electron and that $g = 1$.

APPENDIX II

An Approximate Band System for a ZnS-Cu Electroluminophor with a Cu₂S Surface Phase

As noted above, the process of producing a ZnS-Cu electroluminophor is accompanied by the formation of a second phase, Cu_2S, on the surface of the ZnS-Cu crystals.

The width of the forbidden band in Cu_2S can be estimated from the spectrum of the edge absorption, which has been measured [126]. It is about 1.25 eV. The width of the forbidden band in ZnS is about 3.7 eV.

The position of the Fermi level relative to the bottom of the conductivity band, Δ_F, can be calculated from the equation

$$\Delta_F = kT \ln \frac{N_{\text{eff}}}{N_0}, \tag{A.II.1}$$

where $N_{\text{eff}} \approx 2.5 \times 10^{19}$ cm^{-3}, and N_0 is the equilibrium concentration of electrons in the conductivity band. As already reported in §5 of Chapter II, the equilibrium concentration of electrons in the conductivity band of ZnS monocrystals is about 10^{12} cm^{-3}. In this case, therefore, $(\Delta_F)_{\text{ZnS-Cu}} = 0.45$ eV.

It has been shown [126] that Cu_2S is a semiconductor with hole conductivity and its specific resistance $R_{sp} \sim 10^3$ $\Omega \cdot$cm. The equilibrium concentration of holes in its valence band should be calculable from

$$P_0 = \frac{1}{R_{sp}\mu e}. \tag{A.II.2}$$

Here μ is the mobility of the holes, but its value is unknown.

The mobility in semiconductors varies within wide limits, ranging from about 1 cm^2/V·sec in alkali halides to about 10^3 cm^2/V·sec in elementary semiconductors such as Ge, Si. For calculations based on (A.II.2) we can take values lying between these limits: $P_0 = 6 \times (10^{12}$ to $10^{15})$ cm^{-3}. Since Cu_2S is a hole-type semiconductor, we need to examine the interval between the Fermi level and the top of the valence band. According to equation (A.II.1), this is 0.2 to 0.4 eV. We used a mean value of 0.3 eV (Fig. 30).

Owing to the diffusion of charge carriers from one semiconductor to another, a space charge is produced at the points of contact. This charge creates a field, which leads to equalization of the Fermi levels in these semiconductors and to distortion of the zones in the contact region. The latter effect implies the formation of a potential barrier in this region. The form of this barrier depends on the sign and number of the charge carriers, whose diffusion is responsible for production of the space charge. In the case considered, the potential distribution should be affected by the transition layer of variable composition, formed between the regular crystalline lattices of ZnS-Cu and Cu_2S.

Practically all the contact difference in potential is obviously concentrated in the ZnS-Cu, since the specific resistance of this is greater by several orders of magnitude than that of Cu_2S [100, 126]. The external voltage applied to such a system is also concentrated in the contact region, particularly in the contact region of the ZnS-Cu (Fig. 30b). At a certain voltage U the bottom of the ZnS-Cu conductivity band may drop below the top of the Cu_2S valence band, and tunneling of electrons from Cu_2S into ZnS-Cu becomes possible, as shown by the arrows 1 in Fig. 30b.

APPENDIX III

Calculation of the Transmission Coefficient of the Potential Barrier
at the Boundary between the ZnS-Cu and the Cu$_2$S Surface Phase

According to the theory of the tunneling effect, the probability of penetration of electrons through a potential barrier is proportional to the transmission coefficient of the barrier

$$D = \exp\left\{-\frac{1}{\hbar}\int_0^{L_1} |q(x)|\, dx\right\}, \tag{A. III. 1}$$

where

$$q(x) = \sqrt{2m\Delta(x)}. \tag{A. III. 2}$$

L_1 is the barrier width to the level from which the transition occurs; $\Delta(x)$ is the height of the barrier for this transition at a distance x from the start of the transition.

We will consider this as a unidimensional problem, where x is the distance into the depth of the ZnS-Cu crystal measured from the interface with Cu$_2$S. Then $\Delta(x) = \Delta(0) - eV(x)$, where V(x) is the potential at a distance x from the start of the transition, and $\Delta(0)$ is the barrier height at the start of the transition.

The change of potential in the region of a "depletion" barrier of the Mott-Schottky type follows the equation

$$V(x) = U_0\left[1 - \left(1 - \frac{x}{L}\right)^2\right], \tag{A. III. 3}$$

where L is the width of the region where the difference in potential U$_0$ applied to the crystal is concentrated. This width increases with increasing potential according to the law [see (I. 12)]

$$L = \varkappa_2 \sqrt{U_0}, \tag{A. III. 4}$$

where

$$\varkappa_2 = \left[\frac{\varepsilon}{2\pi e^2 N_C}\right]. \tag{A. III. 5}$$

In this case

$$|q(x)| = \sqrt{2m\Delta(0) - 2eU_0 m\left[1 - \left(1 - \frac{x}{L}\right)^2\right]}. \tag{A. III. 6}$$

$$L_1 = L\left(1 - \sqrt{1 - \frac{\Delta}{eU_0}}\right). \tag{A. III. 7}$$

Integrating $|q(x)|$ over the tunneling transition width L_1, we have

$$\frac{1}{\hbar} \int_0^{L_1} |q(x)|\, dx = \frac{1}{2} \sqrt{\frac{2mU_0}{\hbar}} \left(1 - \frac{\Delta(0)}{eU_0}\right) L\, [y_0 \sqrt{y_0{}^2 - 1} - \text{arch } y_0], \qquad (A.III.8)$$

where

$$y_0 = \sqrt{\frac{1}{1 - \dfrac{\Delta(0)}{eU_0}}}. \qquad (A.III.9)$$

Let us now consider the case when the voltage drop across a ZnS-Cu crystal is large, so that $eU_0 \gg \Delta(0)$. This case is evidently normal, since $\Delta(0) \approx 0.75$ eV, while the voltage used to excite electroluminescence is generally much greater than 0.75 V.

In this case by substituting for y_0, expanding (A.III.8) as a power series in terms of $\dfrac{\Delta(0)}{eU_0}$, and neglecting terms of the second power and above, we obtain

$$\frac{1}{\hbar} \int_0^{L_1} |q(x)|\, dx = \frac{\sqrt{2m}}{3} \frac{\Delta^{3/2}(0)}{eU_0} \frac{L}{\hbar}. \qquad (A.III.10)$$

Then, since the maximum field in the "depletion" barrier of the Mott-Schottky type is

$$E_0 = \frac{2U_0}{L}, \qquad (A.III.11)$$

we have

$$\frac{1}{\hbar} \int_0^{L_1} |q(x)|\, dx = \frac{2\sqrt{2}}{3} \frac{\sqrt{m}\Delta^{3/2}(0)}{e\hbar E_0},$$

so that

$$D = \exp\left\{ -\frac{2\sqrt{2}}{3} \frac{\sqrt{m}\Delta^{3/2}(0)}{e\hbar E_0} \right\}. \qquad (A.III.12)$$

This is analogous to equation (I.1) for the Zener effect.

By using (A.III.4), equation (A.III.10) can, for the case considered, be put in another form,

$$\frac{1}{\hbar} \int_0^{L_1} |q(x)|\, dx = \frac{\sqrt{2m}}{3} \frac{\Delta^{3/2}(0)\, \varkappa_2}{e\hbar \sqrt{U_0}},$$

and equation (A.III.12) can be put in the form

$$D = \exp\left\{ -\frac{\varkappa_2 \sqrt{2m}}{3e\hbar} \frac{\Delta^{3/2}(0)}{\sqrt{U_0}} \right\}, \qquad (A.III.13)$$

i.e., we obtain an equation analogous to (II.26).

LITERATURE CITED

1. O. V. Losev, Telegr. i Telef. 18:61, 1923.
2. G. Destriau, J. chim. phys. 33:620, 1936.
3. G. Wendel, Ann. phys. 12:222, 1953.
4. G. Destriau, J. phys. rad. 14:307, 1953.
5. D. H. Smith, Elec. Eng. 33:164, 1961.
6. H. K. Henisch, Brit. J. Appl. Phys. 12:660, 1961.
7. C. Zener, Proc. Roy. Soc. (London), A, 145:523, 1934.
8. L. H. Hall, J. Bardeen, and F. J. Blatt, Phys. Rev. 95:559, 1954.
9. L. V. Keldysh, Zhur. Eksptl. i Teoret. Fiz. 34:962, 1958.
10. H. J. Fan, Repts. Progr. in Phys. 19:107, 1956.
11. S. I. Pekar, Investigations on the Electronic Theory of Crystals, Moscow-Leningrad, GITTL, 1951.
12. H. Callen, Phys. Rev. 76:1394, 1949.
13. I. T. Nelson and I. C. Irvin, J. Appl. Phys. 30:1847, 1959.
14. R. Newman, W. C. Dash, R. N. Hall, and W. E. Burch, Phys. Rev. 98:1536, 1955.
15. R. Newman, Phys. Rev. 100:700, 1955.
16. D. Rücker, Z. angew. Phys. 10:254, 1958.
17. N. G. Basov, B. D. Osipov, and A. N. Khvoshchev, Zhur. Eksptl. i Teoret. Fiz. 40:1882, 1961.
18. V. A. Uchenkov, Fizika Tverdogo Tela, Collected Papers, Vol. II, p. 209, 1959.
19. L. V. Keldysh, Zhur. Eksptl. i Teoret. Fiz. 37:713, 1959.
20. F. Seitz, Phys. Rev. 76:1376, 1949.
21. B. I. Davydov and I. M. Shmushkevich, Zhur. Eksptl. i Teoret. Fiz. 10:1043, 1940.
22. V. A. Uchenkov and Ch'ên K'o-ming, Fizika Tverdogo Tela 4:3054, 1962.
23. K. W. Böer and U. Kümmel, Ann. phys. 14:341, 1954.
24. K. V. Ber, Izvest. Akad. Nauk SSSR, Ser. Fiz. 24:36, 1960.
25. W. Lehmann, J. Electrochem. Soc. 107:657, 1960.
26. G. Harman, Phys. Rev. 111:27, 1958.
27. K. Maeda, J. Phys. Soc. Japan 13:1352, 1958.
28. P. Zalm, Philips Research Repts. 11:353, 417, 1956.
29. G. Diemer and P. Zalm, Physica 25:232, 1959.
30. W. A. Thornton, Phys. Rev. 116:893, 1959.
31. M. V. Fok, Uspekhi Fiz. Nauk 72:467, 1960.
32. A. Fisher, Z. Physik 149:107, 1957.
33. K. G. McKay, Phys. Rev. 77:816, 1950.
34. A. N. Georgobiani, Optika i Spektroskopiya 12:746, 1962.
35. L. Pauling, The Nature of the Chemical Bond, [Russian translation], Goskhimizdat, 1947.
36. B. V. Nekrasov, A Course of General Chemistry, Goskhimizdat, 1952.
37. J. Weiszburg, Acta Physiol. Acad. Sci. Hung. 11:95, 1960.
38. A. G. Gol'dman, Dokl. Akad. Nauk SSSR, 135:1108, 1960.
39. W. W. Piper and F. F. Williams, Solid State Phys. 6:95, 1958.
40. G. A. Wolff, H. A. Hebert, and J. D. Broder, Phys. Rev. 100:1144, 1955; Halbleiter und Phosphore, Berlin, 1958, p. 547.

41. S. Larach and R. E. Shrader, Phys. Rev. 102:582, 1956.

42. S. Larach and R. E. Shrader, RCA Rev. 20:532, 1959.

43. A. Wachtel, J. Electrochem. Soc. 107:682, 1960.

44. G. A. Wolff, I. Adams, and J. W. Mellichamp, Phys. Rev. 114:1262, 1959.

45. A. Wachtel, J. Electrochem. Soc. 107:199, 1960.

46. H. Lozykowski and H. Meszynska, Bull. acad. polon. sci., ser. sci. mathem., astr. et phys. 8:725, 1960.

47. I. K. Vereshchagin and V. S. Teslyuk, Izvest. Vuzov, Ser. Fiz. 6:114, 1958.

48. A. N. Georgobiani and N. P. Golubeva, Optika i Spektroskopiya 12:802, 1962.

49. R. Frerichs and R. Handy, Phys. Rev. 113:1191, 1959.

50. W. C. van Geel, J. phys. rad. 17:714, 1956.

51. J. N. Bowtell and H. C. Bate, Proc. I. R. E. 44:697, 1956.

52. W. Lehmann, J. Electrochem. Soc. 104:45, 1957.

53. R. Taagepera, R. S. Storey, and K. G. McNeill, Nature 190:994, 1961.

54. W. Lehmann, J. Electrochem. Soc. 105:585, 1958.

55. O. N. Kazankin, F. M. Pekerman, and L. N. Petoshina, Izvest. Akad. Nauk SSSR, Ser. Fiz. 21:721, 742, 1957.

56. I. N. Orlov, Izvest. Akad. Nauk SSSR, Ser. Fiz. 21:731, 1957.

57. I. N. Orlov, V. I. Kas'yanova, and T. F. Uvarova, Collected Annotations of State Union Scientific Research Institutes, 1957, p. 55.

58. A. A. Cherepnev, Optika i Spektroskopiya 2: 770, 1957.

59. V. E. Oranovskii, E. I. Panasyuk, and B. T. Fedyushin, Inzh.-Fiz. Zhur. 2:40, 1959.

60. W. Lehmann, J. Electrochem. Soc. 107:20, 1960.

61. F. J. Sruder and D. A. Cusano, J. Opt. Soc. Am. 45:493, 1955.

62. C. Feldman and M. O'Hara, J. Opt. Soc. Am. 47:300, 1957.

63. N. A. Blasenko and Yu. A. Popkov, Optika i Spektroskopiya 8:81, 1960.

64. N. A. Blasenko, Optika i Spektroskopiya 8:414, 1960.

65. H. C. Froelich, J. Electrochem. Soc. 100:280, 1953.

66. V. N. Favorin, G. S. Kozina, and L. K. Tikhonova, Optika i Spektroskopiya 7:703, 1958.

67. V. N. Favorin and L. P. Poskacheeva, Optika i Spektroskopiya 7:706, 1959.

68. O. N. Kazankin, F. M. Pekerman, and L. N. Petoshina, Optika i Spektroskopiya 7:776, 1959.

69. G. S. Kozina and L. P. Poskacheeva, Optika i Spektroskopiya 8:216, 1959.

70. G. S. Kozina, V. N. Favorin, and I. D. Anisimova, Optika i Spektroskopiya 8:218, 1960.

71. V. N. Favorin and G. S. Kozina, Optika i Spektroskopiya 10:91, 1961.

72. D. W. G. Ballentyne, J. Phys. Chem. Solids 10:242, 1959.

73. G. Destriau, Symposium at Brooklyn Polytechnic Institute, Brooklyn, N. Y., Sept. 1955.

74. P. Goldberg and S. Faria, J. Electrochem. Soc. 107:521, 1960.

75. A. H. Mckeag and E. G. Steward, J. Electrochem. Soc. 104:41, 1957.

76. V. E. Oranovskii and Z. A. Trapeznikova, Optika i Spektsoskopiya 5:302, 1958.

77. A. N. Ince, Proc. Phys. Soc. (London), B, 67:870, 1954.

78. S. Roberts, J. Appl. Phys. 28:262, 1957.

79. N. N. Grigor'ev and Yu. A. Kulyupin, Optika i Spektroskopiya 10:780, 1961.

80. W. A. Thornton, Solid State Physics in Electronics and Telecommunications, Vol. 4, New York, Academic Press, 1960, pp. 2, 658.

81. W. W. Piper and F. F. Williams, Phys. Rev. 87:151, 1952.

82. J. F. Waymouth and F. Bitter, Phys. Rev. 95:941, 1954.

83. W. A. Thornton, Phys. Rev. 102:38, 1956.

84. C. H. Haake, J. Appl. Phys. 28:117, 1957.

85. K. Patek, Czechoslov. J. Phys. B, 10:679, 1960.

86. K. Patek, Czechoslov. J. Phys. 7:584, 1957.

87. F. F. Morehead, J. Electrochem. Soc. 107:281, 1960.

88. A. N. Georgobiani and M. V. Fok, Optika i Spektroskopiya 9:775, 1960.

89. A. N. Georgobiani and Yu. G. Penzin, Optika i Spektroskopiya, Collected Papers on Luminescence, 1:321 1963.

90.   I. S. Marshak, Pribory i Tekh. Eksperim. (5):3, 1957.

91.   M. V. Fok, Optika i Spektroskopiya 2:475, 1957.

92.   L. A. Vinokurov and M. V. Fok, Optika i Spektroskopiya 4:118, 1958.

93.   A. M. Bonch-Bruevich, Ya. É. Karis, and V. A. Molchanov, Optika i Spektroskopiya 11:87, 1961.

94.   T. P. Belikova and M. D. Galanin, Izvest. Akad. Nauk SSSR, Ser. Fiz 25:364, 1961.

95.   M. V. Fok, Uspekhi Fiz. Nauk 75:259, 1961.

96.   V. V. Antonov-Romanovskii, Trudy Fiz. Inst. Akad. Nauk SSSR 2(2-3):157, 1943.

97.   F. A. Kröger and H. I. G. Mayer, Physica 20:1149, 1954.

98.   I. I. Lambe, C. C. Klick, and D. L. Dexter, Phys. Rev. 103:1715, 1956.

99.   F. Abeles and J.-P. Mathieu, Ann. Physik 3:5, 1958.

100. A. Lempicki, D. R. Frankl, and V. A. Brophy, Phys. Rev. 107:1238, 1957.

101. É. I. Adirovich, G. M. Guro, V. F. Kuleshov, and V. A. Uchenkov, Trudy Fiz. Inst. Akad. Nauk SSSR 8:127, 1956.

102. Yu. M. Popov, Optika i Spektroskopiya 6:764, 1959.

103. A. N. Georgobiani and M. V. Fok, Optika i Spektroskopiya 11:93, 1961.

104. A. N. Georgobiani and M. V. Fok, Optika i Spektroskopiya 5:167, 1958.

105. E. E. Bukke, L. A. Vinokurov, and M. V. Fok, Optika i Spektroskopiya 5:172, 1958.

106. L. A. Vinokurov and M. V. Fok, Optika i Spektroskopiya 1:248, 1956.

107. A. M. Bonch-Bruevich and O. S. Marenkov, Optika i Spektroskopiya 8:855, 1960.

108. W. A. Thornton, Phys. Rev. 116:893, 1959.

109. K. Patek, Czechoslov. J. Phys. B, 11:18, 1961.

110. P. Zalm, D. Diemer, and H. A. Klasens, Philips Res. Repts 10:205, 1955.

111. W. W. Piper and F. E. Williams, Brit. J. Appl. Phys. 4:39, 1955.

112. A. N. Georgobiani and M. V. Fok, Optika i Spektroskopiya 10:188, 1961.

113. V. S. Trofimov, Optika i Spektroskopiya 4:113, 1957.

114. W. Lehmann, J. Electrochem. Soc. 107:20, 1960.

115. A. N. Georgobiani, E. Yu. L'vova, and M. V. Fok, Optika i Spektroskopiya (in press).

116. M. V. Fok, Optika i Spektroskopiya 11:98, 1961.

117. G. Destriau and H. F. Ivey, Proc. I. R. E. 43:1911, 1955.

118. J. B. Taylor and G. F. Alfrey, Brit. J. Appl. Phys. Supplement 4:44, 1955.

119. M. V. Fok, Czechoslov. J. Phys., B, 13:99, 1963.

120. E. E. Bukke, L. A. Vinokurov, and M. V. Fok, Inzh.-fiz. Zhur. 7:113, 1958.

121. W. Lehmann, J. Electrochem. Soc. 103:24, 1956.

122. A. N. Georgobiani, E. Yu. L'vova, and M. V. Fok, Optika i Spektroskopiya 13: 564, 1962.

123. W. Lehmann, Illum. Eng. Soc. (N. Y.) 51:684, 1956.

124. P. Lenard, F. Schmidt, and R. Tamaschek, Handbuch der experimentalischer Physik Vol. 2, 1928, p. 23.

125. V. V. Antonov-Romanovskii and I. P. Shchukin, Dokl. Akad. Nauk SSSR 61:445, 1950.

126. L. Eisenmann, Ann phys. 10:129, 1952.

INVESTIGATIONS OF THE CATHODOLUMINESCENCE
OF ZINC SULFIDE
AND CERTAIN OTHER CATHODOLUMINOPHORS

V. L. Levshin, É. Ya. Arapova, A. I. Blazhevich, Yu. V. Voronov,
I. G. Voronova, V. B. Gutan, A. V. Lavrov, Yu. M. Popov,
S. A. Fridman, V. A. Chikhacheva, and V. V. Shchaenko

INTRODUCTION

Despite the importance of cathodoluminescence and its many industrial applications, the basic problems related to the specific method of excitation of crystal phosphors with cathode rays have not been sufficiently investigated.

Cathode excitation exhibits the following properties: 1) nonselective absorption of the energy of cathode rays by the luminophor; almost all absorption takes place in the crystal lattice of the phosphor rather than in the luminescent centers; 2) successive exchange of the energy of the exciting particles, resulting in the appearance of low-energy secondary electrons in the crystal phosphor; 3) luminescence stimulation of localized electrons by the exciting electrons; 4) small penetration depth of cathode rays resulting, on one hand, in an increased importance of the surface layers and, on the other hand, in a much higher excitation density, which usually causes pronounced heating of the active volume of the phosphor.

All these properties of cathode excitation are reflected in the yield and lifetime of the phosphors' luminescence and partially in the emission spectra.

It was the objective of this investigation to examine theoretically and experimentally the problem of the excitation energy losses during cathodoluminescence, establish the approximate magnitude of the limiting cathodoluminescence yield, study the energy exchange of the electron beam during its passage through a luminophor layer, and also to develop individual luminescence processes as functions of the excitation density and temperature. Special attention was given to the study of the inertial properties of ZnS phosphors and their relation to the position and occupation of levels in which electrons and holes are localized.

This study was directed by V. L. Levshin. The theoretical calculations of losses during thermal equalization were performed by Yu. M. Popov (Chapter I). The investigations of the electron energy losses in luminophor layers were carried out by A. I. Blazhevich (Chapter II). The study of the spectral properties and energy yield was made by Yu. V. Voronov (Chapter III). The lifetime and decay laws were investigated by A. I. Blazhevich and V. A. Chikhacheva (Chapter IV). The studies of the occupation of the localization sites of electrons and holes and their relation to the inertial properties were performed by V. B. Gutan (Chapter IV). É. Ya. Arapova developed and prepared the sublimated screens and films (Chapter II). The chemical studies and development and preparation of the luminophors were carried out by I. G. Voronova, A. V. Lavrov, S. A. Fridman, and V. V. Shchaenko (Chapters III and IV).

The authors thank senior designer A. G. Ovchinnikov, radio technicians V. P. Lysov and Yu. A. Platukhin, senior laboratory assistants Z. M. Bruk, S. B. Kondrashkin, N. V. Mitrofanova, L. N. Petrakov, and A. D. Sychkov, and laboratory assistant V. P. Prokhorova for their assistance with the execution of this work.

CHAPTER I

ESTIMATE OF THE ENERGY LOSSES IN CRYSTAL PHOSPHORS
DURING CATHODE EXCITATION
AND THE MAXIMUM CATHODOLUMINESCENCE YIELD

§1.   Concerning Different Types of Losses and the Luminescence Yield during Cathode Excitation

During photoexcitation the luminescence energy yield in the case of crystal phosphors, as well as in the case of liquids, is frequently as high as 0.7-0.8. In the corresponding cases the quantum yield, which does not take into account the energy losses related to the Stokes displacement of the luminescence spectrum, is close to unity.

During cathode excitation we have specific excitation energy losses related to the scattering of primary and secondary electrons. Additional energy losses result from band-to-band transitions induced by the cathode excitation. Upon excitation, holes are formed at different depths in the valence band and the freed electrons, which migrate to the conduction band, initially distribute themselves in a nonequilibrium manner over the levels of the band. Normally, a large fraction of the electrons occupies upper levels of the band. However, thermal stabilization takes place in a very short time (about 10^{-12} sec), i.e., an equilibrium distribution of electrons over the levels of the conduction band and of holes in the valence band is established. During this process an appreciable part of the initial excitation energy is transferred to the crystal lattice in the form of heat.

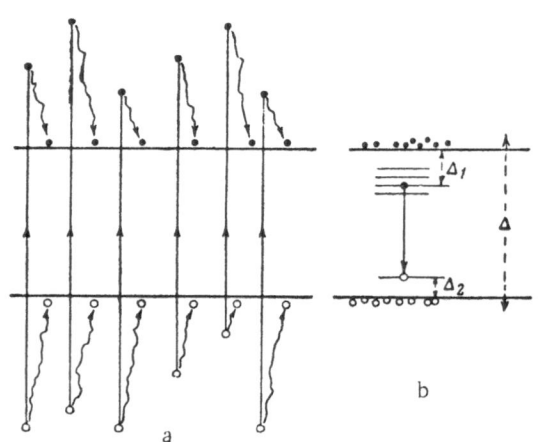

Figure 1a illustrates the thermal equalization process. The initial nonequilibrium distribution of electrons and holes changes during the thermal equalization to their equilibrium distribution near the top of the valence band and near the bottom of the conduction band. The larger the number of electrons reaching the upper levels of the conduction band, or correspondingly the larger the number of holes formed in the lower levels of the valence band, the higher the losses during the thermal equalization. Especially high thermal equalization losses take place, of course, during cathode excitation and during excitation with short-wavelength rays whose quanta are greater than the width of the forbidden band.

Further excitation energy losses take place during the migration of electrons from the equilibrium distribution in the conduction band to luminescent centers whose levels are above the valence band (Fig. 1b). In this case the energy losses are given by the sum ($\Delta_1 + \Delta_2$) of the distance Δ_1 from the average level of the equilibrium distribution of excited electrons in the

Fig. 1. The thermal equalization process of electrons and holes (a) and the origin of the Stokes losses during the luminescence of crystal phosphors (b).

conduction band to the excitation level of the luminescent center and the distance Δ_2 from the level of the center to the average level of the equilibrium distribution of holes in the valence band.

Only these energy losses will we call the Stokes losses.

As in the case of photoexcitation, in addition to the Stokes losses, energy losses also take place during internal and external quenching, which is specially pronounced at high temperatures.

In the following sections we consider the different types of excitation energy losses resulting in lower values of the luminescence yield during cathodoluminescence.

§2. Energy Losses during Electron Scattering

The glow of luminophors may be due to the action of electrons with energies of about 10 eV. However, cathodoluminophors are usually excited with energies between 10 and 30 keV. The ideas presented below pertain to excitation with fast electrons. It is easy to make a qualitative estimate of the losses due to electron scattering.

The electrons which are emitted from the crystal phosphor's surface during electron bombardment and carry away a certain amount of energy may be divided into three groups.

The first, most numerous group consists of very low energy secondary electrons. In some cases the number of such electrons may be greater by one order of magnitude than the number of electrons in the primary beam, so that there are many ionization events per primary electron. The energy of secondary electrons of this type is of the order of 3 to 5 eV, while the energy of primary electrons, as has been stated, is usually a few kiloelectronvolts. For this reason, the energy losses associated with the liberation of slow secondary electrons is much less than 1% of the incident beam energy.

The second group of electrons comprises primary electrons which underwent a number of inelastic collisions and lost a large or small part of their energy through ionization. Upon their emission from a luminophor layer, they may have energies varying between practically zero and an energy close to the energy of the primary electrons incident on the luminophor.

The third group consists of a few primary electrons which, after an elastic collision that does not result in ionization, leave the phosphor's surface with the same velocity as the incident velocity.

These groups of electrons are not separated sharply from each other and they give a continuous energy spectrum. However, the energy distribution of scattered electrons is such that their number decreases with increasing energy. In this case, in the first approximation we can assume that the energy which they carry away is

$$\frac{E_{max}}{2} N,$$

where E_{max} is the energy of the incident electrons and N is the total number of scattered electrons.

It has been shown in [1] and [2] that during electron bombardment of alkali halide crystals with electrons with an energy of 4×10^3 eV the number of elastically and inelastically scattered primary electrons, which depends strongly on the bombarded material and on the angle of incidence of the electron beam on the luminophor's surface, only under most unfavorable conditions exceeds 30% of the number of incident electrons. Thus the excitation energy losses by fast, elastically and inelastically scattered, electrons is not more than 10 to 15%.

It follows from the above discussion that in some cases the energy losses resulting from electron scattering may be lowered to 10-15%. However, it must be stated that these ideas were based on the experiments described in [1, 2] which were performed on alkali halide compounds at a voltage of 4000 V. Similar investigations were not performed on zinc sulfide luminophors at a voltage of 10 to 30 kV.

§3. Energy Losses during Thermal Equalization of Electrons and Holes

During the excitation of crystal phosphors with short-wavelength light or cathode rays the free electrons which are initially produced in the conduction band and the holes in the valence band have a kinetic energy that is many times greater than the energy of electrons and holes thermally equalized by the crystal phosphor's lattice.

Since the lifetime of nonequilibrium carriers in the bands is longer than 10^{-8} sec, they become thermally equalized even before recombination takes place. As a result of the thermal equalization process, excitation energy losses occur.

It will be shown below that, depending on the average energy of the carriers produced in the crystal, the thermal equalization energy losses may be as high as 75%, with the most probable losses being about 60%. It must be emphasized that these losses are independent of the crystal phosphor's purity, depend little on the excitation mode, and do not vary greatly with the specific form of the crystal phosphor. It is natural to assume therefore that such energy losses cannot be eliminated during cathodoluminescence.

The problem of energy exchange of the primary electrons of the cathode beam is very complex, because energy exchange in a solid is a multistage process (through secondary, tertiary, etc., electrons), and is far from being solved. Without considering this problem, we assume that ionization processes are the main processes for all electrons (and holes) in a solid (primary, secondary, etc.), as long as their energy is sufficiently high to transfer an electron from the valence band to the conduction band. Such an assumption is valid, because in the energy range considered (up to 100 keV) the main losses during the penetration of electrons through a solid are connected with ionization [3]. Thus we assume that, as a result of the action of the primary electron beam, electrons with kinetic energies lower than the energy necessary for ionization are produced instantaneously in the conduction band (and holes in the valence band) of the crystal phosphor. * The problem of the energy distribution of these electrons is important in our analysis. However, the energy distribution may be obtained only after solving the problem of the total energy exchange of primary, secondary, etc., electrons (holes).

Fast electrons (and holes) whose kinetic energy is less than a certain value (ionization threshold) cannot form electron-hole pairs. We will determine this energy. Let us assume that we have a fast electron (hole) which is moving in the conduction (valence) band with a quasi-momentum p_0 and a kinetic energy, measured from the bottom of the corresponding band,

$$\varepsilon(p_0) = \frac{p_0^2}{2m_e},$$

where m_e is the effective electron mass. After ionization two additional particles are formed: an electron with a momentum p_{e_2} and a hole with a momentum p_h; the momentum of the original electron after ionization is represented by p_{e_1}. From the laws of conservation of energy and momentum we have

$$p_0 = p_{e_1} + p_{e_2} + p_h; \tag{1}$$

$$\varepsilon(p_0) = \varepsilon(p_{e_1}) + \varepsilon(p_{e_2}) + \varepsilon(p_h) + \Delta, \tag{2}$$

where Δ is the width of the forbidden band.

The ionization threshold is determined from the condition

$$\varepsilon_{ie} \equiv \varepsilon_{\min}(p_0) \equiv \min\{\varepsilon(p_{e_1}) + \varepsilon(p_{e_2}) + \varepsilon(p_h) + \Delta\}. \tag{3}$$

Equating the right side of the equation to zero, we find that at threshold

$$\nabla\varepsilon(p_{e_1}) = \nabla\varepsilon(p_{e_2}) = \nabla\varepsilon(p_h) = v,$$

i.e., the velocities of all the product particles are the same. Using (1) and (2), we obtain

---

* The ionization loss time is shorter than all the remaining characteristic times considered here.

$$\varepsilon_{ie} = \Delta \left(1 + \frac{m_e}{m_e + m_h} \right). \tag{4}$$

If the effective masses of electrons and holes are equal, then it follows from (4) that $\varepsilon_{ie} = \frac{3}{2}\Delta$, i.e., the ionization threshold is one and a half times greater than the width of the forbidden band. Electrons from the conduction band with kinetic energies lower than ε_i cannot take part in the multiplication process. These electrons may collide with the lattice of the crystal phosphor and slow down to the thermal energies, or they may recombine with or without emission and return to the original state.

We will show that the second process is much slower than the first one. Consequently, we can assume that the processes are to a certain extent successive: the first one can reach its completion before the second one develops to a marked extent.

To determine the retardation time of fast electrons (holes), we must find the energy lost by a fast electron per unit time by phonon emission and then find the time (t_r) required to slow it down to an energy kT, determined by the thermal state of the crystal phosphor s lattice:

$$\frac{d\varepsilon}{dt} = - w_p \hbar \omega_p, \tag{5}$$

where t is the time; w_p is the probability of spontaneous emission of a phonon by an electron per unit time; $\hbar \omega_p$ is the energy of the emitted phonon.

Integrating (5), we obtain an expression for the retardation time:

$$t_r = - \int_{\varepsilon_0}^{kT} \frac{d\varepsilon}{w_p \hbar \omega_p} . \tag{6}$$

As is known, fast electrons (with an energy $\varepsilon > \hbar \omega_0$, where ω_0 is the Debye frequency) may upon retardation emit acoustic and optical phonons. This introduces a certain indeterminancy into the value of $\hbar \omega_0$ in (6). However, it has been shown in [4] that, as long as the electrons can emit optical phonons ($\varepsilon > \hbar \omega_0$), the retardation is mainly due to the interaction with the optical lattice vibrations. This result was obtained for valence crystals. Since the majority of cathodoluminophors exhibit to some extent ionic bonds and the cross section for scattering by ionic lattice vibrations in the energy region considered is larger than the cross section for scattering by acoustic phonons, we consider only the scattering by the optical vibrations of the ionic lattice. Even if these cross sections are the same, the optical phonons play the main role in the retardation process, because an optical phonon carries away more energy than an acoustic phonon.

Substituting into (6) the specific form of the probability of scattering by optical vibrations of the ionic lattice from [5], we obtain

$$t_r = \int_{kT}^{\varepsilon_0} d\varepsilon R \varepsilon^{1/2} / \ln \left(\frac{4\varepsilon}{\hbar \omega_0} \right);$$

$$R = \frac{M a^3}{\sqrt{2m\pi} \gamma^2 Z^2 e^4} ; \tag{7}$$

$$M = \frac{M_+ M_-}{M_+ + M_-} ,$$

where $\hbar \omega_0$ is the optical phonon energy; a is the distance between neighboring ions; $M_\pm$ are the masses of the positive and negative ions respectively; Ze is the charge of each ion.

The deviation of γ from unity is determined by the deformation of the electron shells during ionic vibrations. For compounds that are not overly heteropolar γ may be much less than unity. The integral for t_r is not constructed from the elementary functions. However, for estimating purposes (7) may be quite accurately written in the following form:

$$t_r = \frac{2R \left[\varepsilon_0^{3/2} - (kT)^{3/2} \right]}{3 \ln (4\varepsilon/\hbar\omega_0)} . \tag{8}$$

We have replaced the slowly varying logarithm in (7) with a certain average value.

If we estimate the constants characterizing the material ZnS ($Z = 1$; $\gamma = 0.1$; $M = 4 \cdot 10^{-4}$ m_e; $a \approx 3 \cdot 10^{-8}$), with the aid of equation (8), then we find that the retardation from the energy $\varepsilon_0 \approx 3$ eV to kT is achieved in time $t_r \approx 10^{-11}$ sec, i.e., it is much shorter than the emission time of the luminescence quantum ($t \approx 10^{-8}$ sec). Thus a fast electron (hole) becomes thermally equalized before a light quantum is emitted. A superthermal electron (hole) inevitably transfers the kinetic energy acquired during ionization to the lattice. This energy is scattered in the form of heat and comprises a part of the energy losses during cathodoluminescence. To calculate the fraction that these losses represent, we must know the electron (hole) energy spectrum resulting from the ionization processes. As has already been mentioned, this spectrum extends from the bottom of the conduction band $\varepsilon = 0$ to the ionization threshold ε_i. If the energy of all the electrons and holes produced by ionization were close to the ionization threshold (ε_{ie} for electrons and ε_{ih} for holes), the cathodoluminescence energy yield would be minimum. In this case

$$\eta_{min} = 1 - \frac{\varepsilon_{ie} + \varepsilon_{ih}}{\Delta + \varepsilon_{ie} + \varepsilon_{ih}} = \frac{\Delta}{\Delta + \varepsilon_{ie} + \varepsilon_{ih}} . \tag{9}$$

(We do not take into account the losses related to the emission of a quantum with an energy smaller than the width of the forbidden line. This applies to the Stokes losses which occur also during photoluminescence, and will be discussed below.) If in (9) we put $\varepsilon_{ie} = \varepsilon_{ih} = \frac{3}{2}\Delta$, then $\eta_{min} = 0.25$. Thus, when only the losses connected with the thermal equalization of fast electrons and holes are taken into account, the minimum cathodoluminescence energy yield is 25%.

So far, we have considered an unreal case where all the electrons and holes are produced with energies ε_{ie} and ε_{ih}. In the general case, we must take into account the energy distribution of electrons and holes.

Let $f_e(\varepsilon) \varepsilon^{1/2} d\varepsilon$ be the number of conduction electrons* with an energy between ε and $\varepsilon + d\varepsilon$, so that

$$\int_0^{\varepsilon_{ie}} f_e(\varepsilon) \varepsilon^{1/2} d\varepsilon = N_0, \tag{10}$$

where N_0 is the total number of electrons in the conduction band produced by ionization.

There are similar relations also for holes. The cathodoluminescence energy yield is

$$\eta = \frac{\Delta N_0}{\Delta N_0 + \int_0^{\varepsilon_{ie}} \varepsilon f_e(\varepsilon) \varepsilon^{1/2} d\varepsilon + \int_0^{\varepsilon_{ih}} \varepsilon f_h(\varepsilon) \varepsilon^{1/2} d\varepsilon} . \tag{11}$$

As has already been noted above, the energy distribution of electrons and holes produced by ionization is not known and at present there is no method for calculating it quantitatively. However, we know that a distribution for which $\eta = 1$, i.e., where all the electrons are produced at the bottom of the conduction band (holes at the top of the valence band), is not feasible, because the probability of such a process is small (it is proportional to the density of end states, which becomes zero at the edges of the bands). Also, as has been shown earlier, during ionization a large part of the energy of the ionizing particle is inevitably converted into the kinetic energy of the particles produced after the ionization. According to (11), we can determine the energy yield η if we make a simplifying assumption concerning the nature of the distribution $f(\varepsilon)$ of electrons and holes

---

* We assume that energy and momentum are related by the equation $\varepsilon = p^2/2m$, i.e., that the effective mass approximation is valid. In this case the density of levels is proportional to $\sqrt{\varepsilon}$.

resulting from avalanche ionization. We assume that the probability of occupation of any level in the conduction band from the bottom of the band to the energy ε_{ie} is the same, i.e.,

$$f_e(\varepsilon_e) = C_e \quad \text{for} \quad 0 \leqslant \varepsilon_e \leqslant \varepsilon_{ie};$$
$$f_h(\varepsilon_h) = C_h \quad \text{for} \quad 0 \leqslant \varepsilon_h \leqslant \varepsilon_{ih}.$$

This assumption is apparently close to reality, because up till the last stages of ionization the energy of the primary exciting electrons is much greater than ε_{ie} and ε_{ih}. Even if the true distribution function is not a constant, owing to the integral dependence of η on the function f, we do not obtain a numerical value of η that is essentially different from the one calculated earlier. The existing experimental data on secondary emission [6, 7] and on the direct penetration of beams indicate the formation in the solid of electrons with energies a few hundred times higher than the thermal energy (electrons with energies from 1 to 10 eV). Our assumption is in complete agreement with this.

According to the condition (10), we have

$$\left. \begin{array}{l} C_e = {}^3/_2 N_0 \varepsilon_{ie}^{-3/2}; \\ C_h = {}^3/_2 N_0 \varepsilon_{ih}^{-3/2}. \end{array} \right\} \tag{12}$$

After making this assumption, the energy yield in the case of the thermal equalization losses only is determined in the following manner:

$$\eta = \frac{\Delta N_0}{\Delta N_0 + C_e \int\limits_0^{\varepsilon_{ie}} \varepsilon^{3/2} \, d\varepsilon + C_h \int\limits_0^{\varepsilon_{ih}} \varepsilon^{3/2} \, d\varepsilon},$$

whence after some simple calculations we obtain

$$\eta = \frac{1}{1 + {}^3/_5 \left(\dfrac{\varepsilon_{ie} + \varepsilon_{ih}}{\Delta} \right)}. \tag{13}$$

In (13), as was obtained in (4),

$$\left. \begin{array}{l} \varepsilon_{ie} = \Delta \left(1 + \dfrac{m_e}{m_e + m_h} \right), \\ \varepsilon_{ih} = \Delta \left(1 + \dfrac{m_h}{m_e + m_h} \right). \end{array} \right\} \tag{14}$$

Substituting the expressions for ε_{ie} and ε_{ih} from (14) and (13), we obtain

$$\eta = 0.36.$$

Consequently, the cathodoluminescence yield due to the thermal equalization cannot be higher than 36%.

In conclusion it must be stated that the formation of an electron-hole pair by a fast electron with a simultaneous emission or absorption of a phonon is possible. In this case, the lattice, to which the excess momentum is transferred, takes part in the process. The threshold ionization energy is equal to the width of the forbidden band Δ and this increases slightly the cathodoluminescence energy yield. It can be easily calculated that in this case $\varepsilon_{ie} = \varepsilon_{ih} = \Delta$. According to (13), we have

$$\eta = \frac{5}{11} \approx 0.45,$$

The cathodoluminescence energy yield cannot be higher than 45%. However, the probability of such processes is low and the true value of η should be closer to 0.36 than to 0.45.

§4. Stokes Losses during Luminescence of the Crystal Phosphor

It has been stated above that the Stokes losses in crystal phosphors represent losses due to radiation quanta that are small in comparison with the energy width of the forbidden band. This definition differs somewhat from the one used for molecules, where the Stokes losses include usually the thermal equalization losses. It is assumed that the luminescence develops at such a low temperature that practically all electrons in the conduction band are at the bottom of the band and the holes at the top of the valence band at its boundary with the forbidden band. At higher temperatures we must add to the Stokes losses the difference between the average energy of electrons in the equilibrium distribution and the boundary of the conduction band and the difference between the average energy of holes in the equilibrium state and the boundary of the valence band. The coefficients of Stokes losses calculated in such a manner for ZnS-based phosphors are shown in Table 1, where $\eta_{st} = h\nu_{max}/\Delta$ and Δ is the width of the forbidden band.

TABLE 1

Phosphors	ZnS	ZnS-Ag	ZnS-Tu	ZnS-Cu	ZnS-Mn
λ_{max}, mμ	450	430	475	525	585
$h\nu_{max}$, eV	2.77	2.90	2.61	2.38	2.12
η_{st}	0.75	0.78	0.70	0.64	0.57

During the calculations the width of the forbidden band was taken as 3.7 eV. Many authors [8, 9, 10] have assumed that its width reaches 3.9 eV. In this case the losses increase by approximately 5%.

It is evident from the data presented that the Stokes losses in crystal phosphors may be quite appreciable. While in the case of excitation in the fundamental absorption band in dye compounds they usually do not exceed 5%, in crystal phosphors these losses are as high as 25% and in certain cases they may be much higher. The Stokes losses are especially high in phosphors with a wide forbidden band, for example, in alkali halide phosphors in which the width of the forbidden band reaches 9.4 eV (KCl).

§5. Excitation Energy Losses during Energy Migration within a Crystal Phosphor and during Internal Quenching in Luminescent Centers

Losses of this type occur both during cathode- and photoexcitation. The differences in the quenching processes for these types of excitations may be quite large, but not in their principal nature. The applicable excitation energy losses take place during different stages of energy transformation.

In a number of cases, in addition to the luminescent centers, quenching centers are also present in crystal phosphors. When electrons combine with holes localized in quenching centers the latter do not produce any luminescence. The quenching centers may be produced both by special impurities and by other lattice defects.

Frequently, some activators are introduced into the phosphor. In this case it is helpful to introduce the concept of the principal luminescence yield of each group of acting centers. In such multiactivator phosphors individual luminescent centers compete for the use of the activation energy. As the excitation density and temperature are varied in different excitation modes, the excitation energy redistributes itself between individual centers and correspondingly the luminescence yield of one center increases and of another decreases. In this case each of the emitting centers acts as a quenching agent of another competing center.

The energy transfer from one center to another may take place along different paths. At low temperatures resonance transfer is frequently observed, while at higher temperatures the transfer is by displacement of electric charges in the lattice — diffusion of holes from lower trapping levels through the valence band to higher levels. In both cases, owing to secondary reactions, the transformation of one type of luminescence into another is accompanied by losses.

Finally, in the luminescent center itself, owing to a definite relative position of the potential curves of the ground and excited states, the transitions from the excited state to the ground state may be not only radiative, but also nonradiative. The latter correspond to the internal quenching of luminescence.

It follows from this discussion that the complex process of energy transformation of thermally equalized electrons and holes into luminous energy is sometimes accompanied by other processes resulting in the transformation of their energy into heat.

We can decrease all types of the internal and external quenching by reducing the number of various crystal lattice defects which may act as quenching centers.

In practice we encounter various types of phosphors and it happens quite often that the role of radiative transition is very small. However, in better present-day luminophors this type of loss may be minimized under proper excitation conditions.

§ 6. Maximum Energy Efficiency and the Luminescence Yield during Cathode Excitation

The calculation of the photoluminescence energy yield is based on the energy absorbed.

During cathodoluminescence it is usual to take into account all the energy of the beam incident on the phosphor. For this reason, in the latter case the concept of the energy efficiency of luminescence is useful. The energy efficiency is smaller than the energy yield, because it takes into account also the losses associated with reflection of primary electrons and with emission of secondary electrons.

Above we estimated the various types of unavoidable energy losses. If we assume that the quenching processes develop independently of each other (which in the first approximation is undoubtedly true for the main losses), the luminescence energy efficiency may be represented in the form of a product:

$$\eta = \eta_e \eta_t \ \eta_{st} \eta_q.$$

Here η is the luminescence energy efficiency of the phosphor during cathodoluminescence; η_e is the reflection coefficient of primary electrons and secondary emission; η_t is the thermal equalization coefficient; η_{st} is the coefficient of Stokes losses; η_q is the coefficient of external and internal quenching of thermally equalized electrons.

Based on the analysis carried out in §§ 2-5, for the coefficients of the different types of energy losses we may take the following optimum values:

$$\eta_e = 0.9; \ \eta_t = 0.4; \ \eta_{st} = 0.8.$$

We do not take into account the losses associated with the external and internal quenching in the luminescent centers, i.e., $\eta_q = 1$.

Consequently, for the optimum value of the energy efficiency in the most favorable case of the absence of any kind of losses during electron-hole and resonant energy transfer in the phosphor, and in the absence of quenching centers and the internal quenching in the luminescent center itself, we obtain for the optimum energy efficiency of cathodoluminescence the value

$$\eta = \eta_e \eta_t \eta_{st} \eta_q = 0.9 \cdot 0.4 \cdot 0.8 \cdot 1.0 \approx 0.29.$$

In better present-day luminophors $\eta \approx 0.25$. This is almost the limit, because the value 0.29 was obtained under idealized conditions of the development of the process.

Since we assumed the value $\eta_e = 0.9$ for the loss coefficient of the secondary emission, for the maximum energy yield during cathodoluminescence we obtain the value $\eta \approx 0.32$.

CHAPTER II

ELECTRON ENERGY LOSSES DURING BEAM TRANSMISSION
THROUGH A LUMINOPHOR LAYER

§1. Electron Transmission through a Layer of Matter

The understanding of the phenomena which take place when electrons interact with matter depends on the knowledge of the laws governing electron transmission through a layer of matter, their energy loss distribution over depths, and their retardation in solids. A number of papers [11-17] have been devoted to this problem. The experimental methods for determining the penetration and scattering of electrons may be divided into four main groups. Their characteristics are briefly described below.

The method used in [12] was to take microphotographs of the luminescent region of different single crystals during their excitation with electrons having energies between 10 and 40 keV. An expression was obtained for the electron penetration and scattering depth as a function of the initial energy.

The second method consists of measuring the energy and the number of electrons transmitted through matter either with the aid of an electrostatic analyzer [14] or using an electron collector [16, 17]. However, the main drawback of this method is that it necessitates the preparation of mechanically strong, homogeneous, very thin films of the materials investigated. For most crystal phosphors this process is very complicated. For this reason, two other methods are used for investigating luminophors.

In the first one a relation is established between the mean free path of electrons and their energy from the luminescence intensity of a luminescent substrate covered by a layer of the nonluminescent material under investigation. In this manner the electron scattering is determined as a function of the initial energy and penetration depth [13, 15].

The second method consists of preparing thin luminescent films on a nonluminescent substrate [11]. The electron penetration and energy losses in the material are determined from the dependence of the luminescence intensity on the electron energy for different thicknesses of the phosphor layer.

In the papers cited it was confirmed experimentally that the mean free path versus energy relation may be expressed in the form of a power law,

$$R = C\,E^n,$$

where R is the mean free path of electrons in the material; E is the initial electron energy; C is a constant depending on the material under investigation and on the excitation density; n is a constant number which in different papers was assigned different values from 1.3 to 2.

The quadratic dependence corresponding to the Thomson-Widdington law is realized only in a certain electron energy range. Most authors have noted that at energies up to 10 keV the experimental relation deviates from the quadratic law in the direction of smaller n. The reason for this discrepancy has not been established as yet.

To investigate in more detail the cathodoluminescence process in zinc sulfide luminophors, in this paper we attempt to determine the electron beam energy absorbed per unit thickness of the phosphor as a function of the initial electron energy and layer thickness. For this purpose we used the last two methods described above.

§2. Specimen Preparation

The development of the sublimation methods for phosphor preparation opens new ways of investigating the formation of crystal phosphors and their optical properties. Also it widens the practical use of sublimated films for different types of excitation.

The methods of formation of sublimated films may be divided into three main groups.

In the first group we must include the preparation of sublimated films by a gaseous phase reaction followed by condensation of the phosphor on a substrate. Such a method was used, for instance, in [18-20].

The second group includes a two-stage process of film formation consisting of: a) simultaneous evaporation of the basic material and activator on a cold substrate, followed by heating the film in vacuum or a gaseous atmosphere [21-24], and b) successive evaporation of the basic material and activator and heating the film [22-24].

Single-stage processes of direct formation of sublimated films comprise the third group of methods. In this case the basic material and activator are evaporated simultaneously on a substrate heated to the required temperature. The evaporation process is carried out either in vacuum or in a gaseous atmosphere (H_2, HCl, H_2S, or their mixtures).

The film is formed by a direct reaction of the components on a heated substrate [24-29].

In this study we used the latter single-stage method for obtaining sublimated films.

Preparation of an Activatorless Film of Zinc Sulfide with Cubic Structure. The films were formed by evaporating pure zinc sulfide in vacuum. The zinc sulfide was initially treated in a stream of NH_3 at 1000°C and then for one hour in a vacuum at 300°C to remove excess sulfur.

The zinc sulfide was evaporated from an evaporator at 1000°C onto a glass substrate 30 mm in diameter and heated to 400°C. It was shown in [30] that evaporation under such conditions results in the formation of films with a cubic structure. Using this method we prepared zinc sulfide films of different thicknesses from 0.36 to 1.86 μ. They did not exhibit any marked luminescence upon excitation with electrons with energies from 2 to 40 keV. To check this, we prepared a much thicker specimen on a nonluminescent substrate. In this case, the sublimated film also did not luminesce.

Preparation of Sublimated Films of Zinc Sulfide Activated with Manganese. The following single-stage sublimation process is most effective for the ZnS-Mn films.

The composition of the evaporated mixture was 100 parts ZnS, 5 parts $MnCl_2$, 3 parts $MgCl_2$. The mixture was evaporated in vacuum onto a quartz substrate heated to 550°C. Manganese-activated films of different thicknesses may be obtained by this method. The ZnS-Mn films have a cubic structure and exhibit luminescence upon photo- and cathodoexcitation in the region $\lambda_{max} = 590$ mμ.

Film Thickness Determination. The method for determining the film thickness consisted of measuring the displacement of the interference line at the boundary between the surface of the specimen and the surface of the substrate with the aid of an MII-4 interference microscope.

The error in the thickness determination between 0.05 and 0.1 μ was 10 to 15% of the measured value. For measurements of thicker films the error decreased to 3-7%. The thickness measurements were performed after the study of the luminescence characteristics of the films had been completed, because the use of the interference method necessitates drawing of lines on the specimen's surface.

§3. Experimental Study of the Electron Beam Transmission through Sublimated Films

The investigation was carried out with a high-voltage apparatus for measuring the parameters of cathodoluminophors, described briefly in [31] and more fully in Chapter IV of this work.

The luminescence intensity was registered with the aid of photomultipliers FÉU-14A and FÉU-14B and a high-sensitivity microammeter (M-95) capable of measuring currents from 0.002 to 100 μA. The luminescence was measured by transillumination, keeping the photomultiplier in the same position relative to the specimen measured.

The electron beam current density was kept at a constant value of 2×10^{-8} A/cm^2 in all experiments. The current incident on the phosphor was measured with an M-95 microammeter connected to a Faraday cylinder and a specimen holder. Six specimens were investigated simultaneously. The area of the luminous spot was controlled on a standard specimen with the aid of an optical attachment.

The voltage on the accelerating electrode was measured with an electrostatic voltmeter S-96 with voltage ranges 0-7.5, 0-15, and 0-30 kV. From 30 to 40 kV the voltage was measured with the same S-96 voltmeter connected across a resistance voltage divider. The voltage divider was checked for linearity.

During the study of the electron energy losses in nonluminescent ZnS we used glass substrates with detectable luminescence. The substrates were prepared from the same batch of glass. Preliminary measurements allowed us to select pieces of glass with the same luminescent properties over the entire range of electron beam energies investigated. Nonluminescent ZnS films of different thicknesses were deposited by sublimation on the selected glass, 1 mm thick. To prevent the formation of surface charges all the specimens of the sublimated material and also pure glass without ZnS were simultaneously covered with an aluminum film, 0.03 μ thick, by vacuum deposition.

From among a large number of prepared specimens we selected films without any fine holes or scratches. A few groups of specimens were fabricated and the results of the measurements performed on them were in satisfactory agreement. The film thickness at the center of the specimens differed by about 5%.

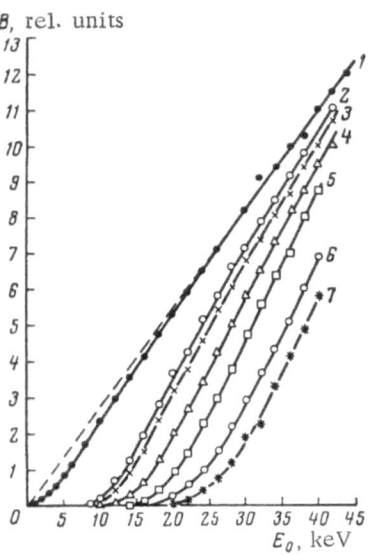

Fig. 2. Dependence of the luminescent brightness of a glass substrate on the initial electron energy for specimens with nonluminescent ZnS layers of different thicknesses. 1) Without ZnS; 2) 0.36 μ; 3) 0.47 μ; 4) 0.7 μ; 5) 1.02 μ; 6) 1.47 μ; 7) 1.86 μ.

Figure 2 shows curves of the dependence of the luminescent brightness of substrates with ZnS layers from 0 to 1.9 μ on the electron energy. The onset of luminescence corresponds to the instant when some electrons penetrate through the surface layer of aluminum and the zinc sulfide film. The absence of luminescence at the beginning of curves 1 and 2 is determined to a great extent by electron absorption in the aluminum layer and also by the "dead potential" of the luminescent substrate. A marked transmission of electrons begins, of course, at energies lower than 1 keV, because even at 1 keV the luminescence intensity is characterized by a 0.03-μA current from the photomultiplier.

The absence of luminescence under the action of low-energy electrons on the remaining curves in Fig. 2 is also determined by complete absorption of these electrons in the ZnS layer. The onset of luminescence indicates a partial penetration of the electrons through the layer.

The curves in Fig. 2 are discussed in greater detail in the following section.

Figure 3 shows the results of similar measurements on specimens of the ZnS-Mn luminophor deposited on a

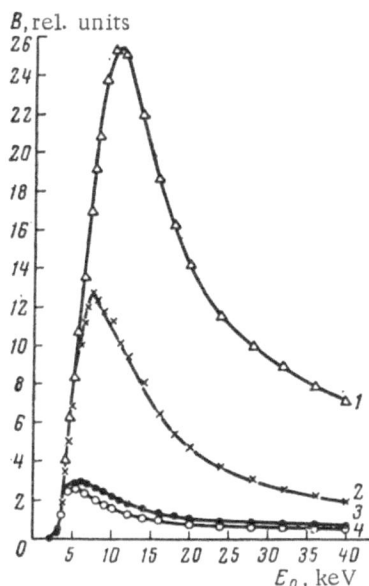

Fig. 3. Dependence of the luminescent brightness of ZnS-Mn luminophor films of different thicknesses on the initial electron energy. 1) 0.3 μ; 2) 0.15 μ; 3) 0.11 μ; 4) 0.09 μ.

nonluminescent quartz substrate.

Our tests have shown, however, that upon electron excitation quartz glass exhibits blue luminescence. For this reason, to isolate the manganese luminescence we used an OG-3 filter, which completely absorbs blue luminescence. In addition to the blue luminescence, quartz glass excited with electrons exhibits also a relatively weak luminescence in the same spectral region as the ZnS-Mn phosphor. A marked luminescence appears at a voltage of the order of 10 kV and it increases with increasing electron energy at a much slower rate than the luminescence of ZnS-Mn. We made the necessary correction for this luminescence.

The fact that the OG-3 filter reduces the luminescent brightness of manganese is not very important, because for the determination of losses it is sufficient to know the relative change in the intensity of ZnS-Mn films of different thicknesses.

The measurements were made on thin films between 0.09 and 0.3 μ thick. Experiments have shown that thicker layers of the ZnS-Mn phosphor obtained by single-stage sublimation do not give good results, because the activator is not distributed uniformly in thicker layers, and this results in large errors in the determination of the electron energy losses. As in the first experiments, all the specimens were covered with an aluminum layer about 0.03 μ thick.

As in the preceding case, the absence of luminescence at energies lower than 2 keV is due to the electron absorption in the aluminum surface layer and also to the "dead potential." The onset of luminescence and the onset of electron transmission through the aluminum layer coincide. The nonlinear increase during the initial stages of luminescence is due to nonlinear energy losses in aluminum; in this energy range the losses in aluminum still constitute a large part of the initial electron energy. A linear increase of the luminescence is observed in the electron energy range in which the absorption of energy in the aluminum layer is negligibly small

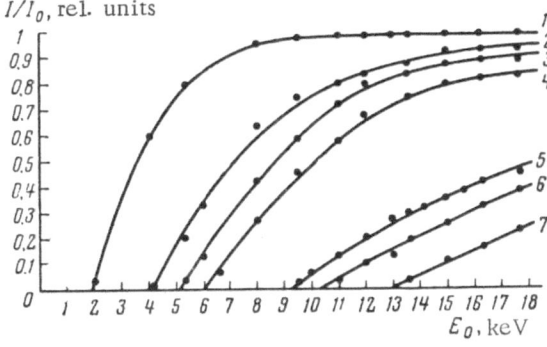

Fig. 4. Curves of the electron transmission through silicon films. Film thicknesses: 1) 0.17 μ; 2) 0.34 μ; 3) 0.42 μ; 4) 0.635 μ; 5) 0.93 μ; 6) 1.44 μ; 7) 1.84 μ.

in comparison with the absorption in the luminophor layer. The deviation from linearity in the region of maximum intensity indicates that electrons have started to penetrate through the luminophor layer. The decay of the curve is due to a reduction in the fraction of the energy transferred by electrons to the phosphor. The thicker the ZnS-Mn film, the larger the displacement of the luminescence peak in the direction of higher energies; the position of the peak is determined by the onset of electron transmission through the luminophor layer.

§4. Determination of the Specific Electron Energy Losses in ZnS

The luminescence intensity is determined by the energy of the electron beam absorbed in the material.

When ZnS layers of different thicknesses, deposited on a luminescent substrate, are irradiated with low-energy electrons, all the electrons are absorbed in the layers and the substrate does not luminesce. During their transmission through the surface layer of ZnS, the electrons impinging on the specimen's surface and having the same initial energy, which is determined by the accelerating voltage, undergo different interactions with the electrons of the material. As a result of these interactions, the electrons from the beam lose their initial monoenergetic nature and are characterized by different transverse mean free paths. The thicker the layer of the material and the lower the initial electron energy, the larger the deviation from the monoenergetic state.

The onset of luminescence corresponds to the beginning of the transmission of a group of electrons which transfer a large part of their energy to the zinc sulfide layer; the remaining energy is transferred to the luminescent substrate. In this case most of the electrons are completely absorbed in the ZnS layer. As the initial energy increases, the number of electrons passing through the film and penetrating the substrate increases and their average energy increases. As the initial energy is further increased almost all the electrons from the primary beam reach the substrate, but with a lower energy than the initial one, because they transfer some of their energy to the layer. The initial electron energy required for the transmission of all electrons through the surface layer depends on the thickness of the film and its physical properties.

The electron absorption process in ZnS may be illustrated qualitatively, using data from [16], in which germanium and silicon films with physical properties similar to those of zinc sulfide were investigated. The film thickness of these materials was the same as the thickness of the ZnS films used in this investigation.

Figure 4 (from [16]) shows curves of the electron transmission through a silicon film. Plotted along the axis of ordinates is the ratio of the electron current through the layer of the material to the primary electron current incident on the surface. The initial electron energy is plotted along the axis of abscissas.

It is evident from Fig. 4 that for an initial electron energy of 18 keV, as the layer thickness is varied between 0.17 and 1.84 μ the number of electrons transmitted through a silicon layer changes from 100 to 27% of the number of electrons on the surface.

To describe completely the interaction of electrons with matter, we must know the energy distribution of electrons at different penetration depths as a function of the initial energy. We did not carry out a study of the energy distribution of electrons in ZnS at different depths, because of great technical difficulties with forming of mechanically strong thin films without a substrate. However, many cathodoluminescence problems may be solved by determining only the energy losses of the entire electron beam in the layer, i.e., the integral losses which take into account both the decrease in the number of electrons per layer thickness and the change in their average energy.

In this study we attempt to determine the specific energy losses of electrons in ZnS for different initial electron energies. In this case, by the specific losses we mean the energy losses of the electron beam calculated per unit thickness of the ZnS layer.

The luminescent brightness of the substrate increases linearly with increasing initial electron energy (curve 1, Fig. 2) for a constant current density. The deviation from the linear law for initial energies up to 25 keV is due to the energy losses in the aluminum surface layer. This was confirmed by additional experiments with luminescent glass covered with aluminum films of different thicknesses. The "dead potential" of the luminescent substrate itself is determined by extrapolating the linear segment of curve 1 into the region of low initial energies; it does not exceed 200 V. The linear segment of curve 1 corresponds to complete electron

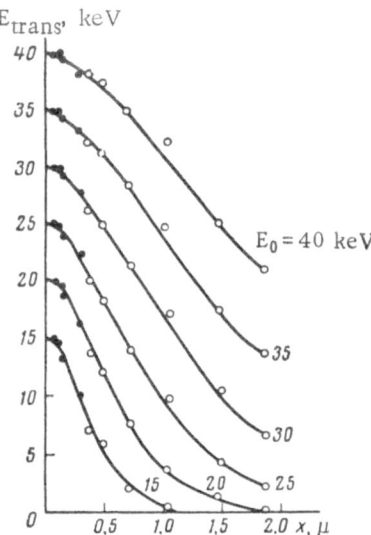

Fig. 5. Dependence of the energy of electrons transmitted through a ZnS layer on the layer depth. The numbers near the curves show the initial electron energy.

transmission with practically no loss of energy in the aluminum layer. In the case of aluminum film thicknesses used in our tests, this was observed for electron energies higher than 25 keV.

The luminescent brightness of the substrate B is proportional to the energy absorbed by it:

$$B = CE_0, \qquad (15)$$

where C is the slope of the linear region; E_0 is the initial electron energy.

Using equation (15), from curves 2-7 in Fig. 2 we can determine the electron beam energy transferred to the substrate after electrons with different initial energies have passed through ZnS layers of different thicknesses. The energy of the electron beam transmitted through a ZnS layer of thickness x is

$$E_{\text{trans}} = \frac{B(x)}{C}, \qquad (16)$$

where $B(x)$ is the luminescent brightness of the specimen's substrate with a surface layer of ZnS of thickness x (for the selected initial energy E_0).

Figure 5 shows the dependence of the energy of the electron beam transmitted right through the ZnS layer on the layer thickness for a few values of the initial energy. In the region of low initial energies we introduced a correction for the energy losses in aluminum. As is evident from Fig. 5, the energy decrease during the penetration of electrons into the layer becomes more abrupt with decreasing initial energy and electrons with low initial energies are completely absorbed by the ZnS layer.

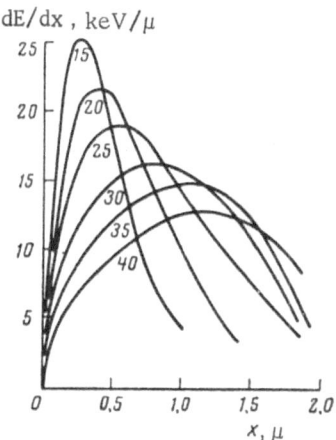

Fig. 6. Distrubution of the specific electron energy losses over the layer depth of zinc sulfide. The numbers near the curves show the initial electron energy.

The energy of electrons transmitted through the layer is defined uniquely by the difference between the initial energy and the energy absorbed in the ZnS layer:

$$E_{\text{trans}} = E_0 - E_{\text{abs}}. \qquad (17)$$

By differentiating the curves $E_{\text{trans}} = f(x)$ with respect to x we can obtain the magnitude and the change in the specific energy losses, taking place during the penetration of the electron beam through a layer x, as functions of the layer thickness for different values of the initial energy. The results of the differentiation are shown in Fig. 6.

As the electrons are slowed down in ZnS, the specific losses first increase as a function of the layer thickness, reach a maximum value and then decrease. The increase of the specific losses in thin layers is due to a gradual slowing down of all the electrons in the beam and the fact that their interaction with the material becomes more intense as their energy decreases. For these thicknesses the number of transmitted electrons decreases slowly with the layer depth. In deeper layers the decrease in the electron energy is accompanied by a rapid decrease in their number due to the absorption in the upper layers. As has been shown in [17], the distribution function of the total transverse mean free paths of electrons passes through a maximum, which in our case corresponds to the peak on the specific losses curves, and then decreases rapidly with increasing layer thickness. As the initial energy decreases, the peak of the specific losses shifts in the direction of smaller thicknesses and the specific losses increase, because as the initial energy decreases electrons are absorbed completely in thinner layers and the energy transfer to the material takes place at a faster rate. Similar experiments were carried out with films of the ZnS-Mn luminophor deposited on a nonluminescent quartz substrate.

From the curves in Fig. 3 we can determine the energy transferred to the quartz substrate through ZnS-Mn films of different thicknesses for a few values of the initial electron energy. The energy of the electron beam transmitted through the phosphor was determined in the following manner. In the region where the electron energy is absorbed completely by the phosphor layer, the luminescent brightness of ZnS-Mn increases linearly with increasing initial energy for a constant current density on the specimen's surface. Using equation (15), from the curves in Fig. 3 we determined the energy absorbed in luminophor layers of different thicknesses for varying initial energy. For a constant initial energy the energy absorbed is determined from the relation

$$E_{abs} \ (x) = \frac{B(x)}{C} \ , \tag{18}$$

where $B(x)$ is the luminescent brightness of a specimen of thickness x; C is the slope of the linear segment of the luminescent brightness curves.

Using equations (17) and (18), we determined the energy E_{trans} transferred to the substrate through luminophor films of different thicknesses for a few values of the initial energy. The energy values obtained are shown in Fig. 5 by dots. They lie on the curve that was obtained for E_{trans} by the first method described above. Thus there is a good agreement between the values of the energy E_{trans} obtained by two different methods.

From the experimental data presented here we can determine the incremental energy absorbed in the layer as a function of the initial electron energy and the most effective layer thickness, from the point of view of utilization of the electron beam energy, for different values of the initial electron energy.

Using equation (17), from the curves in Fig. 2 we determined the energy absorbed in ZnS as function of the initial energy for different thicknesses of the ZnS films (Fig. 7). The 1 keV shift along the axis of abscissas is due to the fact that at lower initial energies all the incident energy is absorbed in aluminum. The linear segment corresponds to the total absorption of the incident energy in ZnS. The slowing down of the growth of

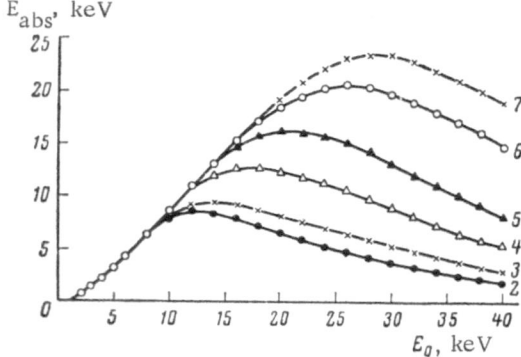

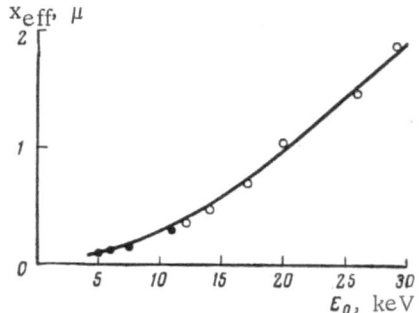

Fig. 7. Dependence of the absorbed energy on the initial electron energy for ZnS films of different thicknesses. 2) 0.36 μ; 3) 0.47 μ; 4) 0.7 μ; 5) 1.02 μ; 6) 1.47 μ; 7) 1.86 μ.

Fig. 8. Dependence of the optimum thickness of zinc sulfide luminophor films on the initial electron energy.

absorption with increasing E_0 is due to the fact that some electrons pass through ZnS and transfer a part of their initial energy to the substrate. The absorption maximum occurs where the transverse mean free path of electrons is approximately equal to the film thickness. As the layer thickness is increased the absorption peak shifts in the direction of higher initial energies and its magnitude increases.

It is obvious then that for a given initial electron energy it is best to use a phosphor layer of such a thickness that the electrons are absorbed completely in the layer. However, it must be noted that in the case of layers of appreciable thicknesses their luminescence intensity is determined not only by the amount of energy absorbed, but also by the absorption of the luminescence produced in the phosphor layer. This is especially important when luminescence is viewed by "transillumination." The values of the optimum film thickness for ZnS-based phosphors as a function of the initial energy (neglecting the luminescence absorption) are shown in Fig. 8. For thin films the dots in Fig. 8 show values of the maximum energy absorbed, determined from the maximum luminescence of the ZnS-Mn luminophor films of different thicknesses from Fig. 3.

It follows from the experiments described above that the specific energy losses of the electron beam during its penetration through a ZnS film increase until a depth corresponding to a certain value of the initial electron energy is reached and then it decreases rapidly (Fig. 6). Thus it becomes possible to determine quite accurately the optimum thickness of luminescent layers for a given voltage.

The point where the maximum electron energy losses occur moves towards the surface of the layer with decreasing initial energy; in this case the specific losses increase and, consequently, the power absorbed per unit volume of the luminophor also increases. As the initial electron energy increases, absorption becomes distributed more uniformly over the layer and the heat release per unit layer thickness decreases slightly.

CHAPTER III

STUDY OF THE LUMINESCENCE, BRIGHTNESS, AND EFFICIENCY SPECTRA OF CATHODOLUMINESCENCE AS FUNCTIONS OF EXCITATION CONDITIONS

After considering the nature of the various excitation energy losses in crystal phosphors during cathode excitation and estimating their magnitude, it is natural to investigate the effect of excitation conditions (current density, voltage, temperature) on the brightness and relative energy yield of luminescence, and establish for the cathode excitation the absolute magnitude of the luminescence energy efficiency in certain typical luminophors relative to the energy of the exciting electron beam.

§1. Apparatus and the Method for Investigating the Brightness, Yield, and Luminescence Spectra

Experimental Conditions. For the purpose of measurements, we used a high-voltage apparatus PRS [32], which allowed us to investigate the brightness and decay of luminescence, the energy yield of luminophors and the spectral composition of radiation. An electron beam provided the excitation. The accelerating voltage was varied from 1 to 35 kV and the current density from 10^{-8} to $5 \cdot 10^{-6}$ A/cm². Measurements were performed both on screens and on luminophor powders at different temperatures. Using two attachments, it was possible to carry out measurement during excitation with ultraviolet light and a cathode beam, either separately or simultaneously. A special attachment was used for investigating luminophor powders "by reflection" [33].

To investigate the luminescence of cathodoluminophors at different temperatures, we used an attachment constructed in FIAN * [34]. It allowed us to investigate powders or screens by reflection. Figure 9 shows the chamber of the apparatus with this attachment. For the electron excitation experiments we used the same electron-optical system as for the work with the standard "by reflection" attachment at room temperature. A special hollow tube filled with nitrogen was used to insert simultaneously four specimens into the packing. The temperature was measured with a calibrated nichrome-constantan thermocouple. Good contact between the specimen's packing and the tube was ensured by the clamping plate. The apparatus was suitable for measurements in the temperature range from -190 to $+150°$C. The two attachment ports permitted simultaneous or separate excitation of the sample with ultraviolet light and an electron beam, and also made it possible to use infrared rays for preliminary irradiation.

Method for Investigating the Brightness of Screens and Powders. The method for measuring the brightness of thick layers and screens was the same, except that separate electron beams were used. The brightness of thick layers was measured from the side of excitation, which was due to an electron beam incident on the phosphor at a 45° angle. The brightness of the screens was measured "by transillumination."

An FÉU-19 photomultiplier connected to a sensitive microammeter M-193 was used for measuring the brightness. The linearity of the spectral characteristics of the FÉU-19 was first checked using Alentsev's photometer [35].

Method for Measuring the Energy Yield in Screens and the Luminescence Efficiency of Cathodoluminophors. The luminescence efficiency of luminophors radiating in the blue-violet region was determined with an

---

* Physics Institute of the Academy of Sciences.

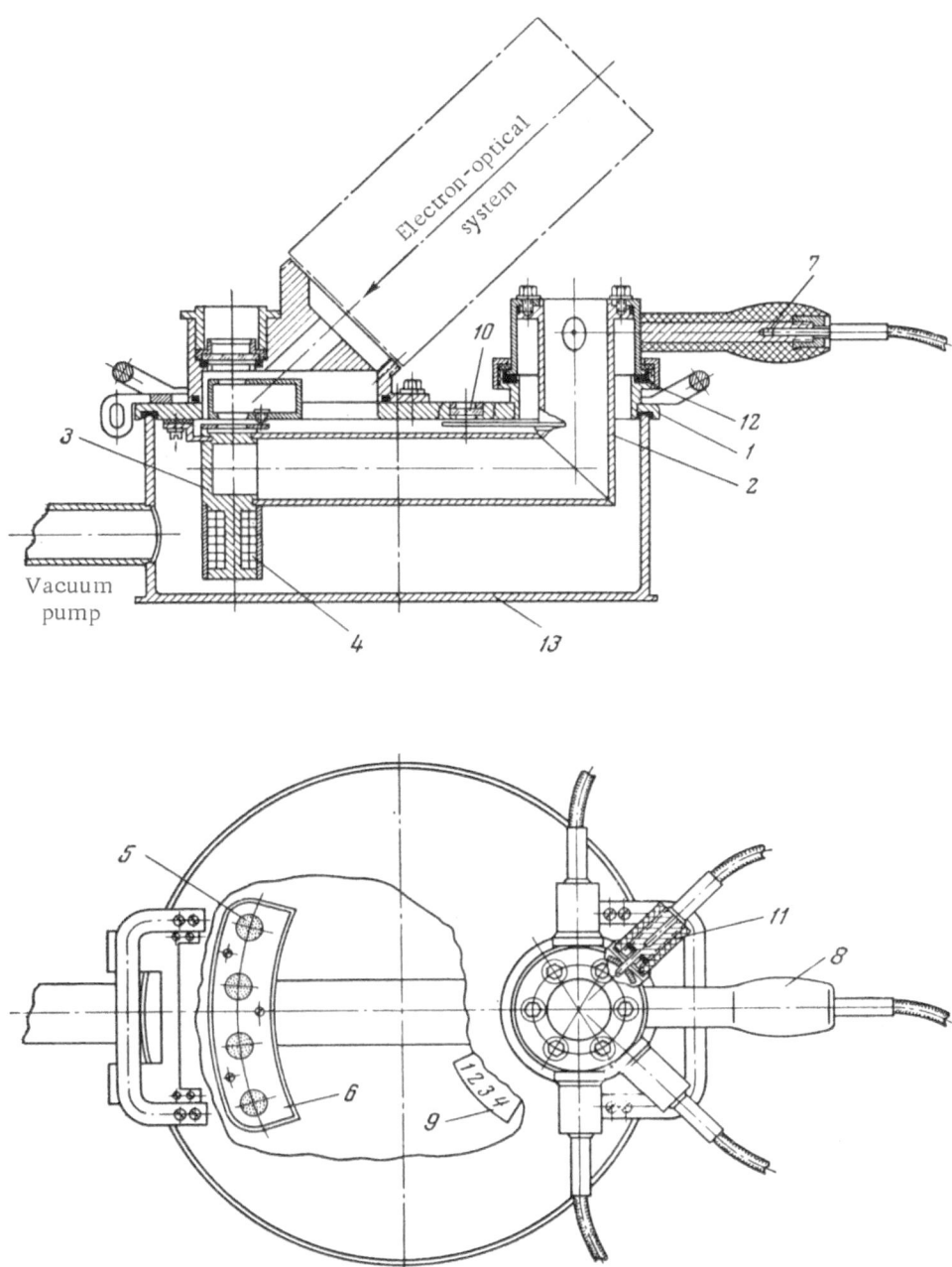

Fig. 9. Chamber of the PRS apparatus with an attachment for investigating cathodo-luminophors at different temperatures. 1) Cover; 2) nitrogen trap; 3) sample holder; 4) electric heater; 5) packing with a luminophor powder; 6) clamping plate; 7) current collector; 8) rotating handle; 9) sample indicator; 10) observation window; 11) vacuum insulator; 12) vacuum seal; 13) chamber of the apparatus.

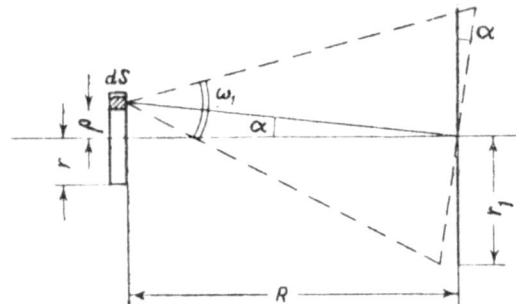

Fig. 10. Scheme for calculating the radiant energy flux from a luminous screen. ds) An element of the screen's surface area; R) distance between the screen and the photocathode; r) radius of the spot (screen); r_1) radius of the FÉU photocathode; ρ) distance from the center of the screen to the element ds.

FÉU-19. The photomultiplier was calibrated in watts with the aid of a standard source — an incandescence lamp SG-2 (6 V, 7.5 W) — with a known energy distribution of radiation. To transform the energy distribution of the comparison source into the energy distribution of the luminophor's radiation and make its intensity comparable, filters ZhS-11 and BG-5 and neutral reducing filters were introduced. The transmission of the light filters was measured with an SF-4. Measurements of the luminous flux from the phosphor and from the standard lamp were made with the aid of the FÉU-19. The data obtained were corrected for the difference in the geometrical position of the comparison source and the luminophor. The energy yield of the luminophor was established by comparing the corrected readings due to the action of the luminophor and the standard lamp.

The energy yield was calculated in the following manner. The filtered radiant flux F from the comparison lamp, incident on the photocathode, was determined from the formula

$$F_{(W)} = \frac{I \int \tau_\lambda \Phi_\lambda d\lambda}{k_0 \int v_\lambda \Phi_\lambda d\lambda} \frac{S_{FÉU}}{l^2}, \qquad (19)$$

where I is the candlepower of the lamp without filters; k_0 is the luminous equivalent (683 lm/W); Φ_λ is the radiant flux power distribution over the wavelengths; τ_λ is the transmissivity of the filters; v_λ is the luminosity of different spectral regions; $S_{FÉU}$ is the illuminated area of the photocathode; l is its distance from the lamp.

The quantity I and the color temperature were obtained from the lamp's specifications; Φ_λ was determined as a relative quantity on the basis of the color temperature; $S_{FÉU}$ and l were determined from the experiment.

The flux from the standard source incident on the photocathode of the photomultiplier caused a deflection of the microammeter's needle. Thus it was possible to calibrate the latter in watts for the spectral composition of radiation corresponding to the phosphor's radiation.

The radiant energy flux from the luminescent screen was calculated in the following manner. The energy flux emitted by the surface area element ds (Fig. 10) of the luminous screen and incident on the photocathode of the photomultiplier is

$$dW = Bds \cos \alpha \, \omega_1, \qquad (20)$$

where B is the energy brightness of the luminescent screen;

$$\cos \alpha = \frac{R}{\sqrt{R^2 + \rho^2}}; \qquad \omega_1 = \frac{S_{FÉU} \cos \alpha}{R^2 + \rho^2};$$

here R is the distance between the screen and the photocathode; ρ is the distance from the center of the screen to the luminescent element.

The radiant flux W transmitted from the luminescent screen onto the photocathode was determined with the aid of the apparatus calibrated in the manner described above. Considering that the linear dimensions of the screen and irradiated spot on the photocathode are commensurable with the distance R between the screen and the photocathode, the radiant flux incident on the photocathode is given by the formula

$$W\ (\text{w}) = \frac{\pi^2 B r^2 r_1^2}{R^2 + r^2 + r_1^2} = N\ (\text{w}),\qquad(21)$$

where B is the energy brightness of the screen, r and r_1 are the radii of the screen and photocathode, N is the power of the radiation incident on the photomultiplier. The latter was determined from the deflection of the microammeter connected to the photomultiplier and calibrated in watts.

Assuming that the screen represents a uniform luminescent surface radiating according to Lambert's law, the flux from the luminescent screen is

$$W_0 = \pi B S_{\text{screen}}.\qquad(22)$$

From equations (21) and (22) we obtain the flux in one direction (by transmission) for the luminescent screen:

$$W_0 = \frac{N S_{\text{screen}}(r^2 + r_1^2 + R^2)}{\pi r^2 r_1^2}.\qquad(23)$$

The energy yield η (W/W) is

$$\eta = \frac{W_0}{VI},\qquad(24)$$

where V is the accelerating voltage and I the beam current.

This method was used for measuring the energy yield of the screens "by transmission" and for measuring the luminescence energy efficiency of powders "by reflection."

<u>Measurement of Spectral Sensitivity of the FÉU.</u> To determine the spectral sensitivity k_λ of the FÉU, we used light sources with overlapping spectral regions. In the 290-400 mμ region the measurements were made with the aid of a hydrogen lamp and a luminophor (a light yellow luminogen) with a constant quantum yield in the 250-450 mμ region. The entry slit of a quartz monochromator was illuminated by a hydrogen lamp. The FÉU was placed in front of the exit slit; the microammeter reading for a given wavelength (P_λ) was proportional to k_λ, the spectral sensitivity of the FÉU at a wavelength λ and a radiation energy E_λ leaving the monochromator,

$$P_\lambda \sim k_\lambda E_\lambda .\qquad(25)$$

Then, instead of the FÉU, the standard luminophor was placed in front of the exit slit of the monochromator. The luminophor's radiation transmitted through a filter, which passed all the luminescent light (GG-14) and stopped the exciting light, was received by the FÉU (reading L_λ),

$$L_\lambda \sim E_\lambda B_{\text{ener}} \sim E_\lambda B_{\text{quant}} \lambda_{\text{exc}},\qquad(26)$$

where B_{ener} and B_{quant} are the energy and quantum yields of the luminophor. Assuming that B_{quant} of the luminogen is independent of λ_{exc}, from the expressions for P_λ and L_λ we obtain

$$k_\lambda \sim \frac{P_\lambda}{L_\lambda} \lambda_{\text{exc}}.\qquad(27)$$

In the 400-600 mμ region the measurements were made with the aid of a monochromator UM-2. An incandescent lamp with a known color temperature (2854°K) and, consequently, with a known energy distribution was placed in front of the entry slit. An FÉU whose photocurrent was measured with a microammeter was placed at the exit.

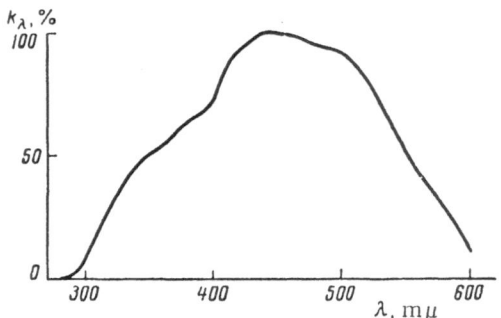

Fig. 11. The spectral sensitivity curve of FÉU-19.

In this case we have the following expression:

$$k_\lambda \sim \frac{A_\lambda}{E_\lambda} D_\lambda , \qquad (28)$$

where A_λ is the reading of the device; E_λ is the energy distribution of the radiation from the incandescent lamp; D_λ is the monochromator's dispersion.

The spectral sensitivity curve of the FÉU-19 is shown in Fig. 11 in relative units.

Method for Measuring the Luminescence Spectra of Luminophors. To measure the luminescence spectra of luminophors, we used a photoelectric method both in the case of photoexcitation and cathodoexcitation.

The flux was transmitted through the window in the chamber to the optical system, which directed it onto the slit of an ISP-51 spectrometer used in conjunction with a photoelectric attachment FÉU-1. After amplification, the current from the photomultiplier was fed to the input of a recording potentiometer ÉPS-157. The true energy distribution in the radiation spectrum of the phosphor was established by dividing the curve recorded by the sensitivity of the equipment.

§ 2. Dependence of the Radiation Spectra of Cathodoluminophors on the Temperature. Comparison with Photoexcitation

Quite frequently, a change in the luminophors' temperature results in a large change in the spectral composition of their radiation, especially if a few activators are present. The effect of temperature on the photoluminescence has been investigated quite thoroughly. The spectral composition changes due to the temperature changes depend on the interaction between activators, the nature of the radiating atom and lattice characteristics. It is difficult to determine the effect of temperature on the cathodoluminescence spectra, because during spectral measurements the phosphor is heated by the incident electron beam.

In our experiments we investigated the luminescence spectra of different classes of luminophors with different activators and coactivators. The tests were carried out on luminophor layers, 0.25 mm thick, on metallic substrates. The phosphor specimens were excited both by cathode rays and light (λ = 365 and 313 mμ) at $-150°$, $+20°$, and $+100°$C.

Phosphor $Sr_3(PO_4)_2$-Eu. As an activator, europium gives different luminescence spectra in different lattices. In the ZnS or ZnS·CdS lattice the luminescence spectrum of Eu is in the form of a wide band in the yellow spectral region, regardless of whether light or electrons are used for activation [36]. As the content of CdS is increased, the luminescence of the activator's band changes, but the peak does not shift. However, at 5% CdS there appears in the spectrum a new band, whose peak shifts in the direction of long wavelengths.

Radiation from both Eu^{2+} and Eu^{3+} may appear in the luminescence spectrum of synthetic single crystals of fluorite [37] activated by the rare earths, depending on the conditions of growth and thermal treatment. A fine structure appears in the luminescence spectrum of Eu^{2+} (CaF_2-Eu prepared under strongly reducing conditions) at the liquid air temperature [38]; the luminescence is violet-blue. The structure of the radiation spectrum of CaF_2-Eu^{3+} preserves it linear character at room and higher temperatures [39]. Red luminescence with a peak at about 640 mμ appears in SrS-Eu. The luminescence spectrum of europium sulfate consists of five bands, with the most intense band at λ = 616 mμ [40].

In thorium-based phosphors [41] the luminescence of Eu is rose-colored and the most intense lines are in the orange and red region of the spectrum.

We investigated strontium orthophosphate activated with europium, $Sr_3(PO_4)_2$-Eu.

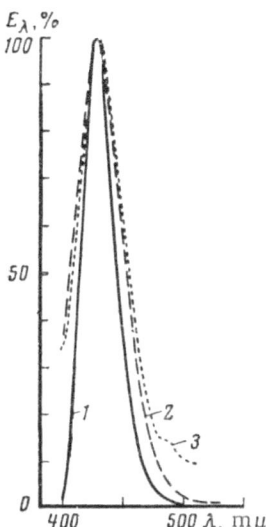

Fig. 12. Luminescence spectrum of $Sr_3(PO_4)_2$-Eu excited with an electron beam (V = 5 kV, j = 10^{-7} A/cm^2) at different temperatures. 1) −150; 2) +20; 3) +100°C.

Fig. 13. Luminescence spectrum of $Sr_3(PO_4)_2$-Eu. 1) Excitation λ = 365 mμ, t = +100°C; 2) excitation λ = 365 mμ, t = −150°C; 3) excitation λ = 313 mμ, t = −150°C (scale I); 4) cathode excitation V = 5 kV, j = 10^{-7} A/cm^2, t = −150°C (scale II).

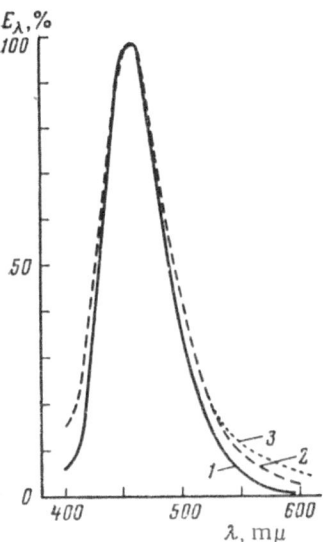

Fig. 14. Luminescence spectrum of the ZnS-Ag luminophor excited with a cathode beam. 1) −150; 2) +20; 3) +100°C.

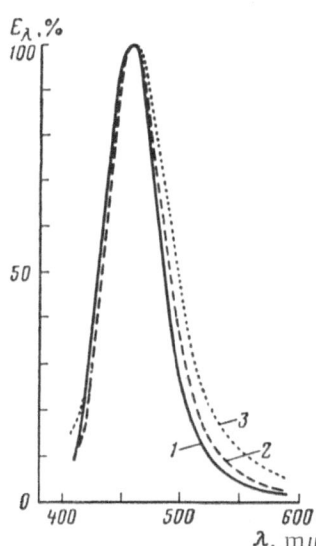

Fig. 15. Luminescence spectrum of the ZnS-Ag luminophor excited with light with λ = 365 mμ. 1) −150; 2) +20; 3) +100°C.

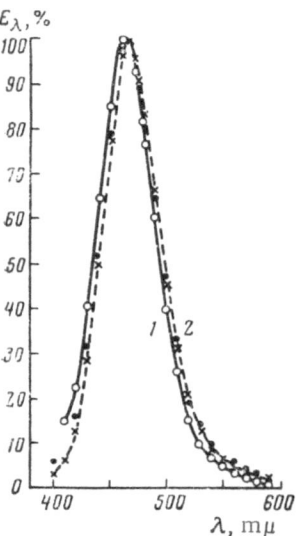

Fig. 16. Luminescence spectrum of the ZnS·CdS-Ag, Al phosphor at a temperature of −150°C. 1) Cathodoexcitation; 2) photoexcitation, λ = 365 and 313 mμ.

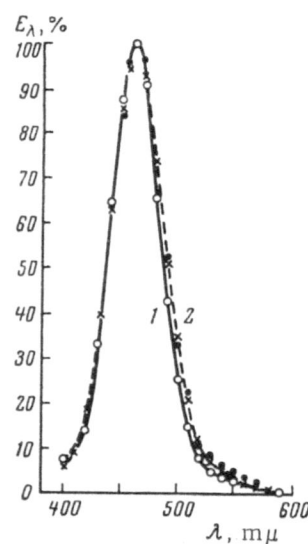

Fig. 17. Luminescence spectrum of the ZnS·CdS-Ag, Al luminophor at a temperature of +20°C. 1) Cathodoexcitation; 2) Photoexcitation, λ = 365 and 313 mμ.

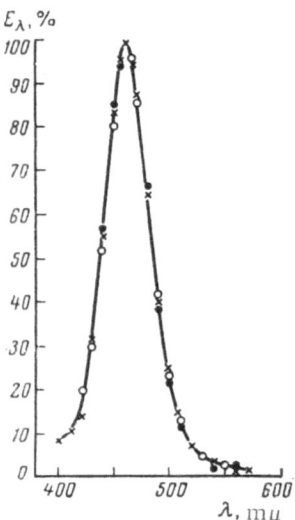

Fig. 18. Luminescence spectrum of the ZnS·CdS-Ag, Lu, Tu luminophor at a temperature of −150°C, independent of excitation.

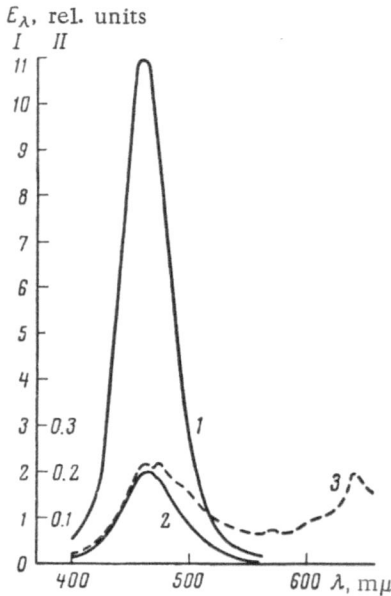

Fig. 19. Luminescence spectrum of ZnS·CdS-Ag, Lu, Tu excited with light λ = 365 mμ. 1) −150 (scale I); 2) +20; 3) +100°C (scale II).

84

The activation of the orthophosphate with europium is considered as an isomorphous substitution of Eu^{2+} for Sr^{2+} [42]. Apparently, at high concentrations of europium Eu^{2+} oxidizes to Eu^{3+} upon annealing in air and this decreases the concentration of centers which produce the blue luminescence.

Figure 12 show the luminescence spectra of $Sr_3(PO_4)_2$-Eu excited with an electron beam (voltage V = 5 kV, current j = 10^{-7} A/cm^2) at different temperatures. At $-150°C$ the luminescence spectrum represents a narrow spectral band; as the luminophor is heated the radiation band becomes wider (the halfwidths are 31, 42, and 46 mμ at -150, $+20$, and $+100°C$ respectively; λ_{max} = 428 mμ) and the luminescent brightness is greatly reduced. The luminescence spectrum of this luminophor excited with light (Fig. 13) differs greatly from the spectrum corresponding to the cathode excitation. The luminescence peak shifts to the ultraviolet spectral region. A number of peaks are observed in the visible region and a peak corresponding to the peak in the cathode beam excitation spectrum is also present. A redistribution of energy among individual bands takes place with increasing temperature (Fig. 13, curve 1).

Phosphor ZnS-Ag. The silver activator is introduced into the batch usually in concentration between 10^{-5} and 10^{-3} g/g. In photoexcited ZnS-Ag phosphors the luminescence of silver is blue.

Figure 14 shows the luminescence spectra of the ZnS-Ag luminophor excited with an electron beam at temperatures between -150 and $+100°C$. The spectrum represents a relatively wide band (the halfwidth at $-150°C$ is 60 mμ) with a peak at 460 mμ that does not shift with increasing temperature. The latter causes a certain broadening of the band in both directions (at $+100°C$ the halfwidth is 68 mμ). When ultraviolet light, λ = 365 mμ, is used for the excitation (Fig. 15) the position of the peak of the radiation band coincides with the position for the electron excitation; the short-wavelength part of the spectrum is also almost the same, but the long-wavelength part at $-150°C$ is somewhat steeper. On the whole, the spectral band is narrower than for the cathode excitation. As the temperature is increased a normal broadening of the spectrum takes place.

The slight difference between the cathodoexcitation and photoexcitation spectra is mainly due to the deeper penetration of light through the phosphor in the latter case. Consequently, the intensity of the short-wavelength part of the spectrum is reduced more than the long-wavelength part, and upon the normalization of the spectra this manifests itself in the development of the long-wavelength branch of the curve. Also, the observed spectral changes frequently may result from the fact that the surface layer heats up during the cathode excitation and the temperature of the acting layer is slightly higher than the average temperature measured during the tests. These factors should to a certain extent also affect the other phosphors.

Phosphor ZnS·CdS-Ag. It was of interest to find out to what extent the temperature properties of the silver band are affected by the introduction of additional ingredients into the composition of the phosphor. It is known that the addition of cadmium sulfide to a phosphor based on zinc sulfide results in a decrease of the width of the forbidden band and a shift of the luminescence spectrum in the direction of longer wavelengths. The ZnS·CdS-Ag luminophor prepared by us contained also Al as a coactivator and Ni, which is a good luminescence quenching agent [43].

The luminescence spectra of the ZnS·CdS-Ag luminophors with the Al coactivator represent a band with a peak at 462 mμ (Fig. 16); they are narrower than the spectra of the ZnS-Ag luminophors (the halfwidth at $-150°C$ is 52 mμ). The spectra excited with a cathode beam and with light (λ = 313 and 365 mμ) are almost the same. As the temperature is increased the luminescence spectrum of the luminophor becomes wider and the luminescence peak shifts into the long-wavelength region of the spectrum, regardless of the wavelength of the exciting light (Fig. 17). The small differences between the photoexcitation and cathodoexcitation spectra may be due to the same causes as in ZnS-Ag.

Phosphor ZnS·CdS-Ag, Lu, Tu. It is known [44] that when rare earth elements are added as coactivators, a strong interaction between silver and the rare earths takes place in ZnS luminophors.

The luminescence spectra of the ZnS·CdS-Ag luminophors with additions of rare earth elements (lutecium and thulium) represent essentially the luminescence band of the main activator (silver). At high temperatures a luminescence peak of thulium is superimposed on its background. The spectra are exactly the same for all the types of excitation used and the thulium luminescence is as yet not observed at $-150°C$ (Fig. 18). As the temperature is increased, the spectrum becomes wider and the thulium luminescence appears. In the case of the electron beam excitation its brightness passes through a maximum and at +100°C the luminescence becomes weaker. In the case of excitation with ultraviolet light (λ = 365 mμ), in addition to the main peak, associated with silver, and the thulium peak (λ = 476 mμ), two new thulium peaks appear as the temperature is raised to +100°C. There is a weak peak in the yellow region and a strong luminescence peak in the red region of almost the same intensity as the main peak (Fig. 19). The luminescence at temperatures above zero is much weaker in comparison with the luminescence at -150°C (scale II). It must be emphasized that the peak intensity in the red region in the ZnS-Tu luminophor without Ag and Lu is very low in comparison with the main luminescence peak (λ = 476 mμ). Consequently, the observed development of this peak in ZnS·CdS-Ag, Lu, Tu at a higher temperature represents a well-defined sensitization of the red luminescence of thulium by silver or lutecium. This sensitization is apparently connected with the redistribution of holes among activator levels [45].

Phosphor ZnS-Tu. Finally, we investigated the luminescence spectra of ZnS luminophors activated with thulium alone. It has been found in [46] that the photoluminescence of ZnS-Tu luminophors is blue and exhibits a short afterglow. We prepared a cathodoluminophor with a thulium concentration of 5×10^{-5} g/g. Figure 20 shows its luminescence spectra at -150, $+20$, and $+100°C$. It is evident from the figure that the luminescence spectrum of Tu at $-150°C$ exhibits a line structure and is much narrower than the spectra at temperatures above $0°C$ (λ_{max} = 476 and 478 mμ). As the temperature is increased, each narrow band becomes wider and, consequently, the sharp structure of individual lines vanishes. The luminescence spectrum of thulium combines into a relatively narrow band over the interval 465-480 mμ with a peak at 476 mμ. The form of the spectrum is the same for electron beam excitation and photoexcitation. The red peak of the Tu luminescence at + 100°C is not observed in phosphors activated with thulium alone.

The experimental investigation of the luminescence spectra of various phosphors as a function of the temperature showed that the spectra of certain, mainly multiactivator, phosphors undergo appreciable changes. Also, in some cases cathode excitation was observed to produce unique properties.

§3. Spectral Composition of Radiation as a Function of Excitation Conditions

In phosphors with one activator the spectral composition of radiation is one of the stable properties of luminescence. It has been observed [47, 48, 49] that during photoluminescence the spectral composition of radiation does not vary with the excitation power. In complex phosphors changes in the excitation intensity may result in large changes of the luminescence spectrum due to an interaction between activators.

We investigated the dependence of the spectral composition of radiation on the accelerating voltage and on the current density for a few luminophor classes (phosphates, ZnS, and ZnS·CdS phosphors). The excitation was provided by a constant electron beam and the luminophor powders were measured "by reflection." The results of our measurements of the luminescence spectra of the luminophors as a function of the accelerating voltage for a constant current density are shown in Fig. 21.

It follows from Fig. 21 that the variation of the accelerating voltage is not accompanied by a shift of the peak in excess of the experimental error. The experimental points lie on one curve and do not exhibit any ordered deviation when the energy of the bombarding electrons is varied between 5 and 20 keV. The results of

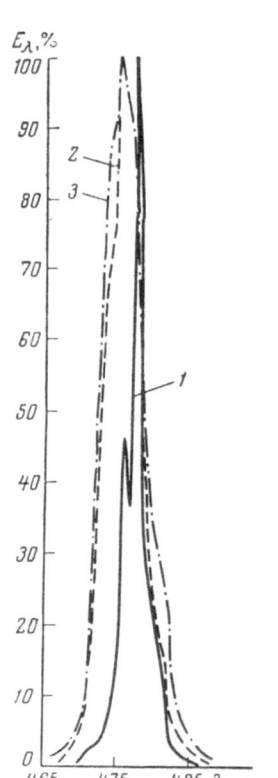

Fig. 20. Luminescence spectrum of the ZnS-Tu luminophor for cathodo-excitation. 1) -150; 2) $+20$; 3) $+100°C$.

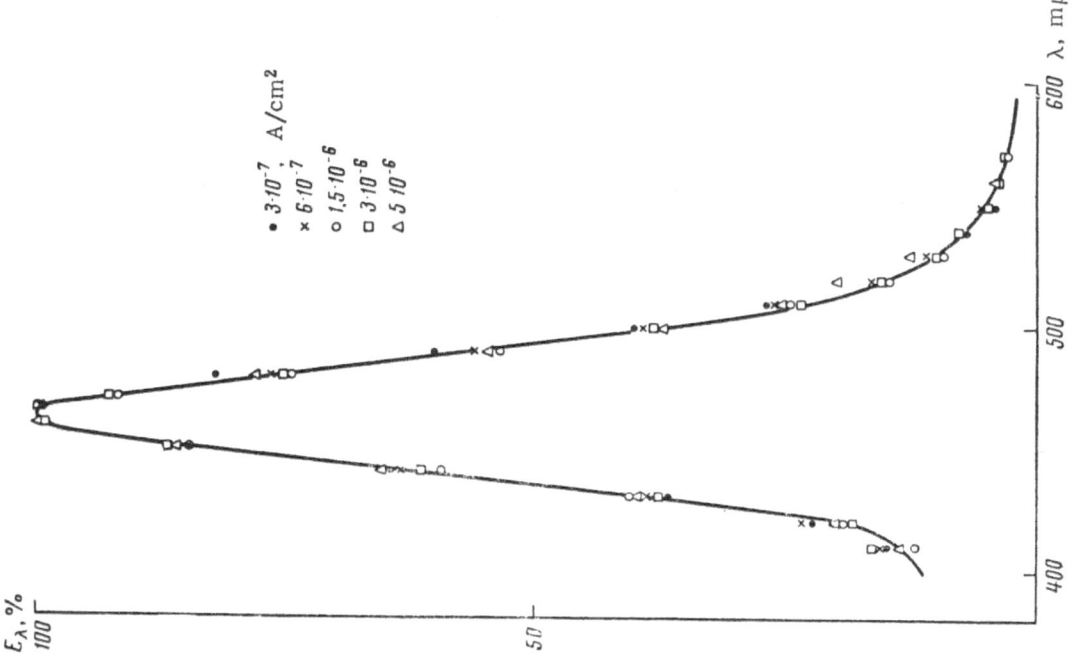

Fig. 22. Independence of the luminescence spectra of the ZnS·CdS-Ag, Al phosphor of the current density. Voltage 5 kV.

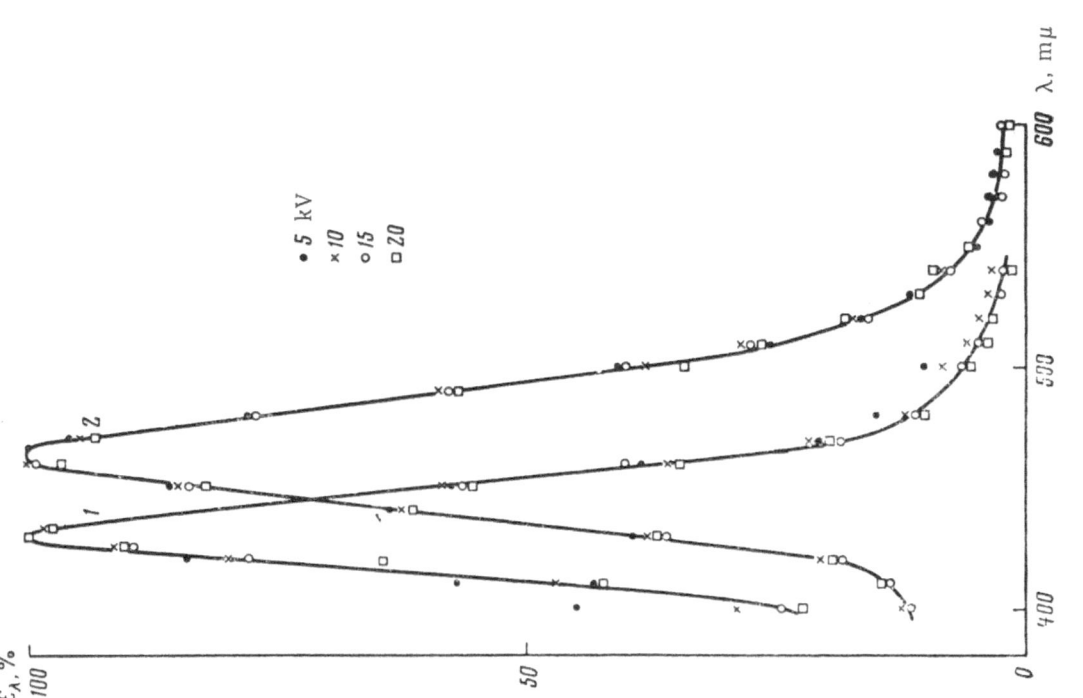

Fig. 21. Independence of the luminescence spectra of Sr₃(PO₄)₂-Eu (1) and of ZnS·CdS-Ag, Al (2) of the accelerating voltage. Current density 10^{-7} A/cm².

our measurements of the luminescence spectra as a function of the current density, shown in Fig. 22, indicate that in this case the peak also does not shift.

It is possible that the small variations of the decay part of the spectral band in strontium orthophosphate observed at an accelerating voltage of 5 kV are connected with the experimental error, because the intensity of the luminescence spectrum was quite low in this region.

§ 4. Dependence of the Cathodoluminescent and Photoluminescent Brightness on the Temperature

The luminescent brightness of phosphors for a constant excitation depends strongly on the temperature. Moreover, the intensity variation of even a single band in phosphors of complex chemical composition may follow a very complex law [50-52].

Heating to a few hundred degrees results normally in the reduction of the intensity of all luminescence bands, because of the development of temperature quenching. However, in the case of cathodoluminescence the quenching takes place much later [53]. This is due to the development of short-duration processes.

The complex temperature dependence at lower temperatures is connected with the temperature-produced redistribution of the excitation energy between the activators and the quenching agents.

We investigated the luminescence brightness of a few luminophor powders "by reflection" (layer thickness 0.25 mm) as a function of the temperature. The measurements were made with the aid of a special attachment to the PRS equipment, constructed in FIAN [34], using light (λ = 365 and 313 mμ) and an electron beam (V = 5 kV, j = 10^{-7} A/cm^2) for excitation. Figure 23 shows the brightness curves as a function of the temperature for various phosphors excited with light.

In all cases, the intensity begins to decrease at −150°C, but the curves are not monotonic. In a number of temperature intervals a slight increase in the brightness takes place and it is followed by a subsequent decrease. A rapid decay begins at +50°C.

The shape of the curves depends on the composition of the phosphor. The ZnS-Ag phosphor is most stable and ZnS-Tu, Ni is least stable. The latter has the most complex brightness versus temperature curve.

The excitation with the shortest wavelength of light (λ = 313 mμ) has almost no effect on the temperature dependence of brightness. The order in which the measurements are made also has little effect on the brightness variation. The shape of the curves is essentially independent of whether the luminophor is heated from low to high temperatures or cooled from +150 to −150°C.

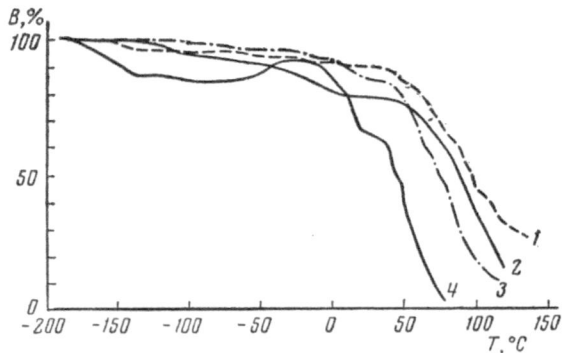

Fig. 23. Temperature dependence of the luminescent brightness in the case of photoexcitation (λ = 365 mμ). 1) ZnS-Ag; 2) ZnS·CdS-Ag, Al; 3) ZnS·CdS-Ag, Lu, Tu; 4) ZnS-Tu, Ni.

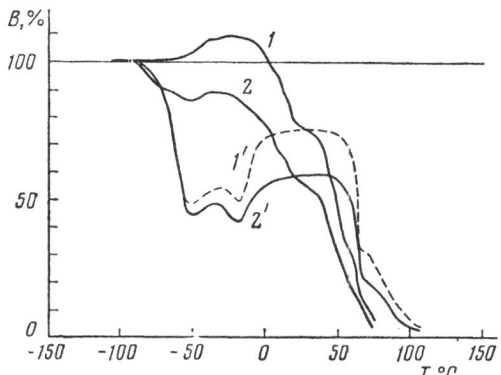

Fig. 24. Dependence of the luminescent brightness of two bands of the ZnS-Tu, Ni luminophor on the temperature. 1) λ = 476 mμ; 2) λ = 478 mμ for photoexcitation (λ = 365 mμ); 1' and 2') the same for cathodoexcitation.

The temperature behavior of the luminescence intensity is quite different for photoexcitation and for excitation with cathode rays.

Figure 24 shows the temperature variation of the intensity for two components of the complex blue band of the ZnS-Tu, Ni phosphor (Fig. 20) isolated with an ISP-51 spectrograph.

Both these components vary in an almost identical manner, but the curves corresponding to the photo- and cathodoexcitation are quite different.

In the range from −50 to 0°C and near +40°C luminescence intensity peaks are observed for both types of excitation, except that in the case of photoexcitation the peak in the region −50 to 0°C is approximately 1.6 times as high as at +40°C, while for the cathodoexcitation it is 0.8 of the luminescence at +40°C.

An even greater difference in the temperature dependence of the intensity of photo- and cathodoluminescence was observed in the complex ZnS·CdS-Ag, Al, Ni phosphor (Fig. 25). In the case of photoexcitation the brightness decreases almost monotonically as the temperature is increased from −100 to +100°C.

As is evident from Fig. 25, the shape of the cathodoluminescence curves varies and it depends greatly on the previous tests made on the specimens. A study of the temperature dependence of brightness during a cooling cycle after the same dependence had been determined in the phosphor excited with λ = 313 mμ gave curve 1, a study during heating of the phosphor excited with λ = 313 mμ gave curve 2, and a study of a fresh specimen curve 3. It appears that the cathode excitation results in unstable changes of the operating part of the phosphor and this is reflected in changes of its emissivity.

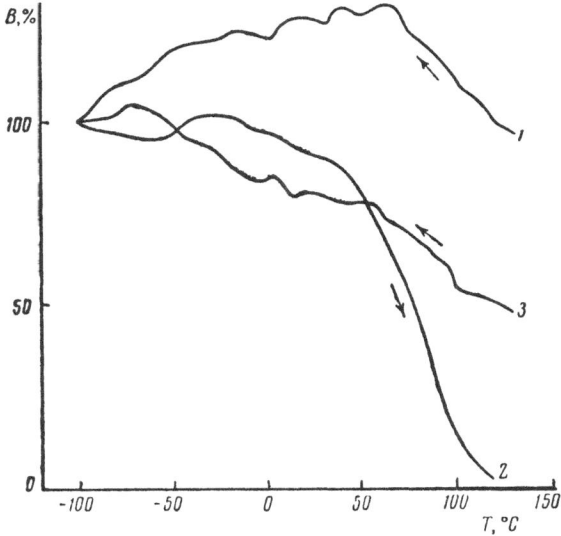

Fig. 25. Dependence of the luminescent brightness of ZnS·CdS-Ag, Al, Ni luminophor on the temperature. 1,3) Cathodoexcitation; 2) photoexcitation (λ = 313 mμ).

As the same time, the increase of the cathodoluminescence intensity with increasing temperature may be due to the liberation of the residual gases which condense on the luminophor's surface. This explanation was proposed by the authors of [54]. As yet, we cannot explain the interesting features of the temperature variation of brightness that are observed during cathodoluminescence, as additional data are required.

§5. Dependence of the Luminescent Brightness on the Excitation Current Density

We would expect that, when a luminophor is excited with a monoenergetic electron beam, the increase in the brightness would be proportional to the number of incident electrons. This theoretical hypothesis holds for quite a wide range of current densities [55, 56], in some cases up to 10^{-4} A/cm^2. However, at very high excitation densities we begin to observe a deviation from this proportionality — the brightness increases at a slower rate. This deviation is observed sooner in phosphors with a long-duration afterglow.

Under certain conditions the brightness may cease to increase and may even decrease [57, 58]. In the region of increasing brightness its variation with increasing current density follows a complex law. The current saturation effect has not been as yet fully explained, even though it was analyzed theoretically in [59, 60]. We enumerate the following main causes of the reduction in the rate of increase of brightness.

1. The charge accumulated on the phosphor results in a decelerating electric field whose intensity increases with the beam current.

2. The surface of the screen is heated during the electron bombardment and temperature quenching becomes important.

3. The effect of kinematic saturation of luminescent centers [59] connected with a finite lifetime of the excited state and a small number of the luminescent centers at the penetration depth of the electron beam.

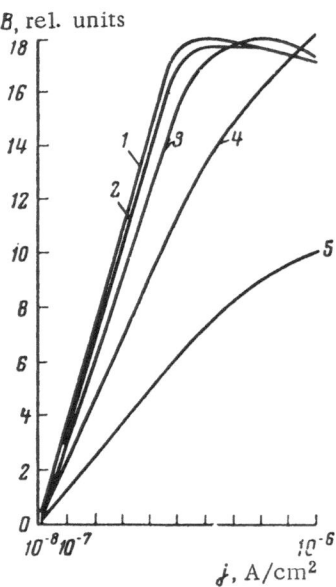

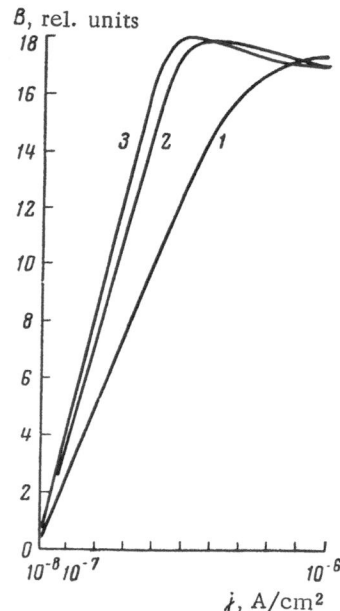

Fig. 26. Dependence of the luminescent brightness of screens on the current density for V = 15 kV. 1) ZnS-Ag; 2) ZnS·CdS-Ag, Lu; 3) ZnS·CdS-Ag, Al; 4) ZnS-Tu; 5) Sr$_3$(PO$_4$)$_2$-Eu.

Fig. 27. Dependence of the brightness of the ZnS-Ag luminophor on the current density for different accelerating voltages. 1) 10 kV; 2) 15 kV; 3) 20 kV.

To determine the luminescence efficiency of typical phosphors and screens we measured the variation of their luminescent brightness with the current density. Figure 26 shows the variation of brightness as a function of the current density for aluminized screens. Initially, all the curves rise linearly, then the rise rate decreases appreciably, a luminescent brightness saturation is reached, and finally a slight decrease takes place. The region of the linear increase of the brightness with the current density is important for the determination of the maximum luminescence efficiency of the luminophors. We used this region in the following.

The rate at which the brightness saturation is reached is different for different phosphors. In our experiments we observed a distinct relation between the saturation rate and the decay rate. Such luminophors as ZnS-Tu and strontium orthophosphate activated with Eu, which exhibit a certain afterglow (of the order of a few milliseconds), tend to saturate at much higher current densities than ZnS-Ag, in which the afterglow is of long duration. It follows from this that the exhaustion of all possible excited centers plays a definite role; a faster deexcitation increases the cycle of the centers and this is equivalent to increasing the number of centers that can be excited and can take part in radiation per unit time. In this case higher current densities are, of course, needed to exhaust the free centers and reach the limiting luminescent brightness. However, the reduction of the brightness rise rate with increasing current density is most affected by the charge accumulated in the luminophor layer near the glass substrate [61] and the creation of an opposing electric field, which results in a deceleration of the incident electrons and heating of the screen.

Figure 27 shows the dependence of the luminescent brightness of a ZnS-Ag luminophor screen on the current density for different values of the accelerating voltage. It is evident from the figure that at low densities of the excitation current (from 10^{-8} to $3-5 \times 10^{-7}$ A/cm^2), at which the screen is far from being saturated, the brightness increases with increasing voltage for each current density, because the fast electrons penetrate deeper and excite additional centers. At higher current densities, however, an increase in the voltage does not result in a brightness rise. In this case even a small number of electrons that reach deeper layers at a low voltage are sufficient to excite strongly these deep layers. Nothing is gained by increasing the electron velocity in this case.

Comparing the curves B = $f(j)$ (Fig. 27) obtained at different voltages, it is evident that, as the accelerating voltage is increased, the luminescence peak shifts in the direction of lower electron beam current densities. If we calculate the power of the electron beam W = Vj for the three luminescence peaks, we observe that the luminescent brightness saturation in all cases occurs at almost the same electron beam power per unit area of the phosphor:

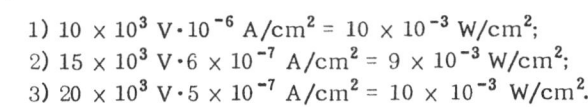

1) 10×10^3 V·10^{-6} A/cm^2 = 10×10^{-3} W/cm^2;
2) 15×10^3 V·6×10^{-7} A/cm^2 = 9×10^{-3} W/cm^2;
3) 20×10^3 V·5×10^{-7} A/cm^2 = 10×10^{-3} W/cm^2.

However, at V = 10^4 V the electrons do not penetrate through the luminophor layer; conversely, at V = 2×10^4 V a large number of electrons penetrate through the layer and carry away up to 20% of the energy. For this reason, the excitation density should be higher in the first case than in the second.

Figure 28 shows curves of the dependence of the luminescent brightness on the current density for thick layers of different luminophors. As in the case of the screens, initially the curves rise linearly with increasing current density; then the rate of the brightness rise becomes slower. It should be noted that here the rapid decrease of the rise rate takes place much before the onset of the brightness saturation in the screens (Fig. 27). This effect is connected with the formation of space charges which slow down the

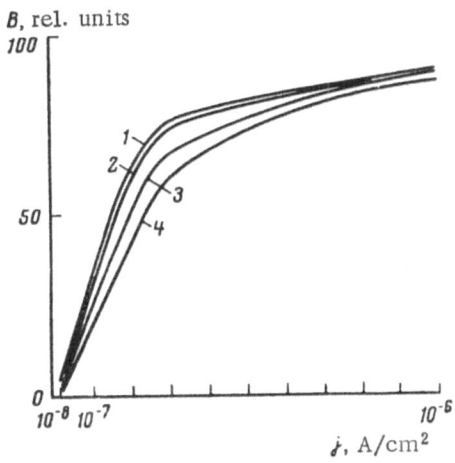

Fig. 28. Dependence of the luminescent brightness on the current density for thick layers of luminophors. 1) ZnS-Ag; 2) ZnS·CdS-Ag, Lu; 3) ZnS·CdS-Ag, Al; 4) ZnS-Tu.

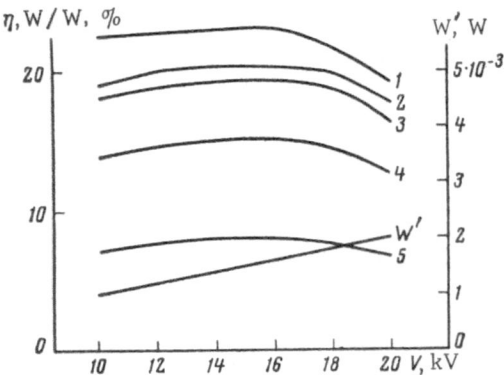

Fig. 29. Dependence of the energy yield of screens on the accelerating voltage. Current density 10^{-7} A/cm². 1) ZnS-Ag; 2) ZnS·CdS-Ag, Lu; 3) ZnS·CdS-Ag, Al; 4) ZnS-Tu; 5) $Sr_3(PO_4)_2$-Eu. W' is the electron beam power.

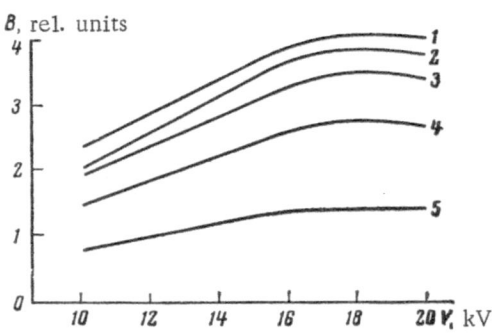

Fig. 30. Dependence of the luminescent brightness of screens on the accelerating voltage. Current density 10^{-7} A/cm². 1) ZnS-Ag; 2) ZnS·CdS-Ag, Lu; 3) ZnS·CdS-Ag, Al; 4) ZnS-Tu; 5) $Sr_3(PO_4)_2$-Eu.

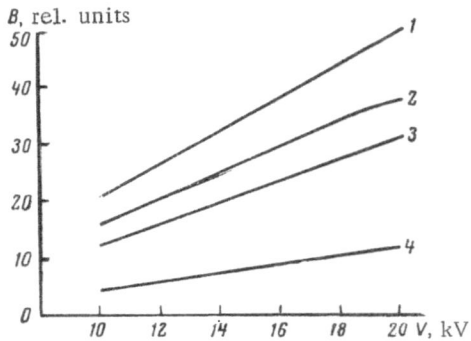

Fig. 31. Dependence of the luminescent brightness of thick luminophor layers on the accelerating voltage. Current density 10^{-7} A/cm². 1) ZnS-Ag; 2) ZnS·CdS-Ag, Al; 3) ZnS-Tu; 4) $Sr_3(PO_4)_2$-Eu.

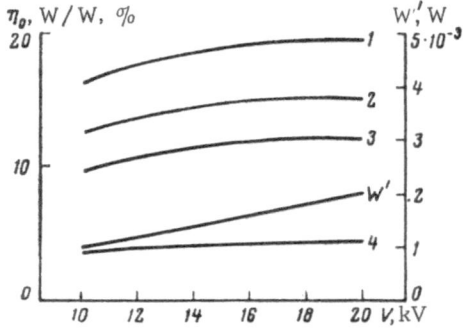

Fig. 32. Dependence of the energy efficiency of the luminescence of powders on the electron accelerating voltage. 1) ZnS-Ag; 2) ZnS·CdS-Ag, Al; 3) ZnS-Tu; 4) $Sr_3(PO_4)_2$-Eu. W' is the electron beam power.

excitation electrons at the luminophor's surface. The formation of such a field around thick luminophor layers is quite normal because their conductivity is extremely low.

§6. Determination of the Energy Efficiency of the Cathodoluminophors' Luminescence

As has been shown in Chapter I, owing to the secondary electron emission and a partial scattering of the primary electrons by the luminophor's surface, it is difficult to determine the electron beam energy absorbed by the phosphor during cathodoluminescence. For this reason, the ratio of the energy of the luminescence emitted to the total energy of the electron beam incident on the phosphor is used to characterize the efficiency of the phosphor. In Chapter I we called this ratio the energy efficiency of the cathodoluminophor's luminescence.

The luminescence and energy yields are used in industry to characterize the quality of a luminescent screen relative to its brightness during cathode excitation. These parameters are determined by measuring

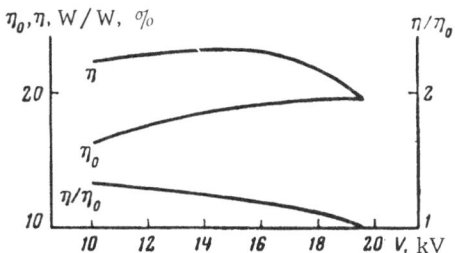

Fig. 33. Dependence of the energy efficiency and yield of a luminophor on the accelerating voltage (η in an aluminized screen, η_0 in a thick layer of powder).

directly the luminous flux or energy emitted only from the front of the screen. In the case of thin nonaluminized screens, the luminous flux emitted by the screen on the excitation side and the absorption losses incurred during repeated scattering of the luminescence within the screen are neglected. Thus the energy yield is the ratio of the energy emitted by the screen (from the front) to the energy supplied at the instant of excitation. The energy yield describes a given luminescent screen. In addition to the luminophor's properties, it is affected by (1) unidirectional observation of the luminescence, (2) energy losses within the screen, and (3) penetration of the bombarding electrons through the screen in the case of thin screens and high-energy electrons. All these losses of the luminous and excitation energies must be taken into account during the determination of the energy efficiency of the luminophor's luminescence, which is a characteristic of the material.

A number of papers [62-65] were devoted to the problem of measuring the luminescence energy yield during cathode excitation. However, owing to differences in the excitation conditions and methods for measuring the electrical quantities, the results of individual authors differ greatly, even for the same phosphor classes.

§7. Dependence of the Energy Yield and Luminescence Efficiency on the Accelerating Voltage

We consider the dependence of the luminescence efficiency of a luminophor,

$$\eta = \frac{\int I_\nu d\nu}{Vj},$$

i.e., the ratio of the luminous energy $\int I_\nu d\nu$ emitted per unit time to the energy supplied by the electron beam, on the accelerating voltage. The penetration depth of the exciting electrons ($R \cong CV^2$) increases with the accelerating voltage, i.e., the working volume of the luminophor increases. Consequently, the excitation density per unit volume is

$$\frac{jV}{R(V)} = \frac{j}{V},$$

where j is the electron current density, $R(V)$ is the transverse mean free path of primary electrons, and V is the accelerating voltage.

Thus the excitation density per unit volume decreases with increasing accelerating voltage as $1/V$. For lower volume densities the luminescence energy efficiency should increase somewhat, because of the reduction in heating and nonradiative processes. This is in agreement with the shape of the energy yield curves for screens shown in Fig. 29. It is evident from these curves that the energy yield increases slightly with increasing accelerating voltage. Upon further increase of the voltage the luminescence energy yield of the screens reaches a broad peak and then begins to decrease. This decrease is due to the fact that electrons penetrate through the luminophor layer on the screen and some of their energy is transferred to the substrate. To check this conclusion, we compared the variation of the luminescent brightness of screens and thick layers of luminophors as a function of increasing electron excitation energy (Figs. 30 and 31).

Inasmuch as in the thick layer we observed a monotonic linear increase of the luminescent brightness with increasing voltage, in the screens a brightness peak is reached at 15 kV and then a slow decrease of the brightness takes place. This effect is undoubtedly connected with the penetration of some of the excitation electrons through the screen (see Chapter II).

The curves of the dependence of the luminescence energy efficiency of powders (η_0) on the electron accelerating voltage in Fig. 32 show that the luminescence efficiency increases slightly with increasing V. Apparently, this is due to the fact that the relative amount of energy carried away by secondary electrons decreases with increasing accelerating voltage and the surface layer of the phosphor is heated to a lesser extent.

Figure 33 shows the dependence of the energy efficiency of the ZnS-Ag luminophor in an aluminized screen (η) and in a thick layer of luminophor powder (η_0) on the accelerating voltage. The thickness of the luminophor layer on the screen was 1.1 mg/cm^2 and the thickness of the luminophor powder layer was 1 mm. It is evident from the figure that the ratio η / η_0 decreases with increasing accelerating voltage. This is due to the reduction of absorption of the cathode rays in the screens with increasing voltage, as compared with the energy absorption in a thick layer of the powder.

§8. Dependence of the Energy Yield and Luminescence Efficiency on the Current Density

As has been shown above, a gradual saturation of the luminophors' brightness takes place with increasing current density of the electron beam. Such a variation of the brightness with the current density indicates that the energy yield and luminescence efficiency of the phosphors decrease with increasing current density. The decrease of the luminescent brightness and energy yield at higher current densities is due mainly to the increase of the decelerating electric field produced by the negative electronic charges on the surface and in the volume of the luminophor, and to the increase of the temperature of the working layer. Below we consider the variation of the energy yield of screens and of the luminescence efficiency of powders as a function of the current density.

The dependence of the luminescence energy yield of various phosphors on the current density for aluminized screens is shown in Fig. 34 and the luminescence efficiency of thick luminophor layers investigated "by reflection" is shown in Fig. 35.

The energy yield of the screens is constant for low current densities and then it decreases smoothly with increasing current density. In the case of the powders the efficiency begins to decrease with increasing current density at the lowest current density used, because of the space charge accumulation. At $j = 2 \times 10^{-7}$ A/cm^2 the decrease becomes quite appreciable. The decrease of the luminescence efficiency of the powder layers is apparently due to the small working volume at the voltages used. For an accelerating voltage of 10 kV the mean free path of the primary electrons in the luminophor is only 10^{-4} cm and therefore the excitation density per unit volume is quite high. Consequently, the space charges may accumulate and the phosphor heats up.

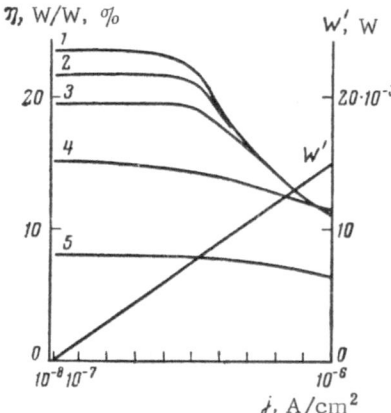

Fig. 34. Dependence of the luminescence energy yield of screens made of different phosphors on the current density. 1) ZnS-Ag; 2) ZnS·CdS-Ag, Lu; 3) ZnS·CdS-Ag, Al; 4) ZnS-Tu; 5) Sr$_3$(PO$_4$)$_2$-Eu. W' is the electron beam power.

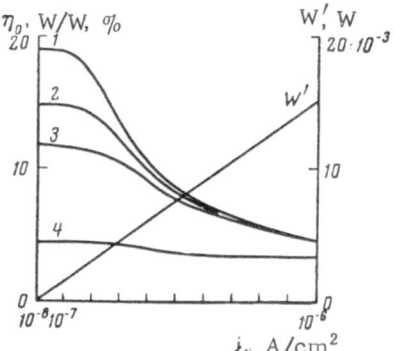

Fig. 35. Dependence of the luminescence efficiency of thick luminophor layers on the current density. 1) ZnS-Ag; 2) ZnS·CdS-Ag, Al; 3) ZnS-Tu; 4) Sr$_3$(PO$_4$)$_2$-Eu. W' is the electron beam power.

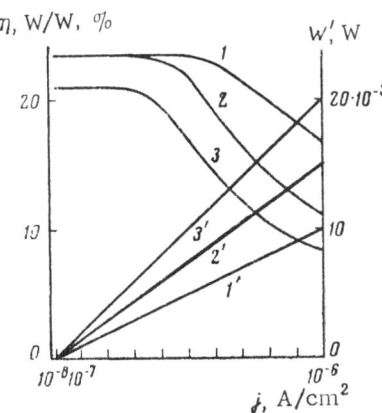

Fig. 36. Dependence of the energy yield of the ZnS-Ag luminophor on the current density for different values of the accelerating voltage. 1) 10 kV; 2) 15 kV; 3) 20 kV. 1', 2', 3') Power variation at the corresponding voltages.

Figure 36 shows the dependence of the energy yield of an aluminized screen made of the ZnS-Ag phosphor on the current density for different values of the accelerating voltage. The maximum luminescence energy yield of the ZnS-Ag luminophor for accelerating voltages of 10 and 15 kV is approximately the same for current densities up to 3 $\times 10^{-7}$ A/cm^2. It is equal to 23.3 W/W (%).

For an accelerating voltage of 20 kV the maximum energy yield is much lower than the energy yield observed for lower voltages (curve 3, Fig. 36). This is related to the penetration of high-energy electrons through the entire thickness of the screen and the transfer of a certain amount of energy to the nonluminescent substrate.

At high current densities (of the order of 10^{-6} A/cm^2) the luminescence brightness of the screen remains constant, regardless of the accelerating voltage (Fig. 37). The increase in the power supplied upon the transition from 10 to 20 kV is compensated for by the increasing transmission of high-energy electrons through the screen and, possibly, by slightly greater heating of the phosphor. The energy yield decreases correspondingly (curve 2, Fig. 37). The fact that the energy yield remains optimum for high current densities in aluminized screens as compared with thick layers of powder shows that in the latter case the reduction in the luminescence efficiency is connected not with the nature of the luminophor, but with the development of secondary processes, such as the formation of a decelerating field of residual charges in the luminophor layer and the possible heating of the layer. Consequently, under optimum excitation conditions we can expect a decrease of the luminescence efficiency at high current densities. In thick layers the luminescence efficiency of luminophors is approximately 20% lower than in screens.

§9. Absolute Energy Efficiency of Luminescence in the ZnS-Ag Cathodoluminophor

In §§6-8 we considered the dependence of the energy yield of screens and the luminescence efficiency observed in powders on the excitation conditions (variation of V and j). To determine the true luminescence efficiency (η_a), the value of η must be corrected for the energy losses in the screen associated with light emitted from within the screen or the layer outwards and, in the case of screens, also for the energy carried away by electrons transmitted through the screen.

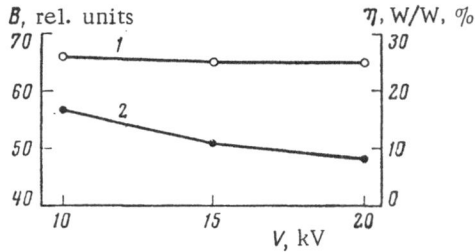

Fig. 37. Dependence of the brightness and energy yield on the accelerating voltage. Current density 10^{-6} A/cm^2. 1) Brightness; 2) energy yield.

To determine the energy efficiency of a luminophor in the form of a screen, the screen thickness must be such (1 to 2 mm) that the incident electron beam is totally absorbed. To take into account the luminescence propagated in the direction of the gun, the screen is coated with an aluminum layer, which has a large coefficient of reflection in the screen's radiation region. Under these conditions, after the determination of the losses associated with the reflection of light from the aluminum layer, we calculate the total luminous energy flux of the cathodoluminophor's luminescence due to the energy of the electron beam incident on the phosphor. Dividing it by the electron beam power supplied to the screen, we obtain the energy efficiency of the luminophor's luminescence. In the case of thick luminophor layers, to the experimental data obtained on

TABLE 2

Current density, A/cm^2	Brightness, rel. units		
	nonaluminized screen	aluminized screen	powder
$1 \cdot 10^{-7}$	26	63	52
$2 \cdot 10^{-7}$	56	156	128
$3 \cdot 10^{-7}$	92	240	192

the luminescence energy efficiency we should apply a correction, taking into account the losses associated with the luminescent light emitted from within the luminophor.

As has been pointed out during the development of the calculating method in §1, in this case the main role is played by the absorption of the luminescent light which at first travels inwards in the luminophor and only then reverses its direction, because of scattering in the infinitely thick layer. For the light initially directed towards the surface the losses are not taken into account, because the excited layer is assumed to be infinitely thin. Despite the fact that such a calculation is artificial and approximate, it gives satisfactory results [66, 67].

This calculation does not take into account the effects associated with the heating of the layer and the formation of the decelerating field. To determine the luminescence efficiency of a luminophor, we must use that range of current densities in which the luminescent brightness is proportional to the current density and the energy yield is constant. In our case this region is from 0.1 to 5×10^{-7} A/cm^2.

To calculate the efficiency, we used the values of brightness obtained at a voltage of 15 kV, because the above-described experiments on the determination of the dependence of the luminescent brightness on the accelerating voltage have shown that at 15 kV the energy absorption is distributed over the entire layer of the luminophor on the screen, while the losses due to the penetration of electrons through the screen are still not appreciable. To find out the effect of aluminizing the screens, we compared the luminescent brightness of nonaluminized screens operating "by transmission," of aluminized screens, and of thick layers observed from the excitation direction. The results are shown in Table 2.

It is evident from the table that, despite certain losses associated with the reflection of light from the aluminum coating, aluminizing increases the luminescent brightness by a factor greater than two. This is explained by the abrupt reduction of the amount of charge on the luminophor and reduced heating of the phosphor. The same reasons can be used to explain the reduced brightness of a thick phosphor layer, which is almost exactly equal to twice the brightness of a nonaluminized layer operating "by transmission." It should be noted that the powder layer was excited at an angle of 45° and, consequently, the number of reflected electrons was increased. At 5 and 10 kV the brightness of the aluminized screen is equal or even smaller than twice the brightness of the nonaluminzed screen operating "by transmission." The reducing role of the aluminum coating becomes quite pronounced at these voltages.

We will calculate the true energy efficiency of the ZnS-Ag luminophor in an aluminized screen and in a thick powder layer. The experimental value of the energy yield for an aluminized screen gives

$$\eta_a = \frac{W_{obs.rad.}}{Vj},$$

where $W_{obs.rad.}$ is the power of the observed radiation. By definition, the effect of the decelerating field of the screen's charge and the losses associated with the reflection of electrons by the luminophor are not taken into account in the calculation of the luminescence efficiency. However, the reflection and absorption of the luminescent light by the aluminum layer must be taken into account. These losses decrease $W_{obs.rad.}$ by 5% [66, 67]. Also, we must take into account the inactive absorption in the bond of the screen, which is as high as 2%, and 2% for the reflection on the glass substrate.

As has been shown above (p. 96), the optimum energy yield obtained experimentally was 23.3 W/W (%). Taking into account these variations of $W_{obs.rad.}$, we obtain for the luminescence efficiency of the phosphor in screens η_a = 25.6 W/W (%). This is in good agreement with the theoretical value of the optimum efficiency (Chapter I). To find the luminescence yield, we must take into account also the excitation energy losses associated with the retardation of electrons by the screen's charge and with the reflection and scattering of primary and secondary electrons. As has been shown in Chapter I, the value of the luminescence yield under the optimum excitation conditions is higher by 10-15% than the value of efficiency, i.e., it may be as high as 30%.

In the case of powder layers, the incident radiation was at an angle of 45° also for the thick layers (1 mm). The optimum experimental value of the luminophor's luminescence efficiency was $\eta_0 = 18.9$ W/W (%). However, this value does not take into account the absorption of light emerging from within the phosphor. According to [67], the true luminescence efficiency $\eta_{a,0}$ is approximately equal to $\eta_0/0.91$. Thus in our case

$$\eta_{a,0} = 20.8 \text{ W/W } (\%).$$

The reasons for the lower value of the luminescence efficiency of luminophors in thick layers as compared with the luminescence in aluminized screens have been discussed above.

CHAPTER IV

INVESTIGATION OF THE DURATION AND THE DECAY LAW OF CATHODOLUMINESCENCE AND THEIR RELATION TO THE POSITION AND OCCUPATION OF LOCALIZATION SITES OF ELECTRONS AND HOLES

§1.   Luminescence Decay and the Concept of the Duration of Luminescence

The luminescence decay is determined by the kinetics of the de-excitation process and therefore it gives important information about its nature. In the simplest cases it leads to a unique conclusion. Thus the de-excitation of a system of discrete centers gives an exponential decay, while a simple recombination process results in a second-order hyperbolic decay.

In crystal phosphors the decay of luminescence is caused by a complex process of recombination of electrons freed from their traps at different depths with holes localized in various interacting luminescent centers. In the cases where during excitation a certain amount of energy is absorbed by luminescent centers, the decay of the directly excited centers is superimposed on this process.

The liberation of electrons from their traps at a certain depth at a given temperature progresses exponentially, but branching of the system of localized energy levels and broadening of levels of a certain type result in a superposition of many exponentials; the decay is further complicated by the secondary trapping of electrons and the interaction between various radiation centers. Consequently, the recombination luminescence decay in crystal phosphors is of a complex form, which is described approximately by E. Becquerel's very flexible empirical formula

$$I = (a + bt)^{-\alpha} = A\,(a_1 + t)^{-\alpha}.$$

The physical significance of the constants in this formula is obvious, but they are not easily expressable in terms of the constants characterizing the luminophor.

The solution of practical problems is very much dependent on the knowledge of the luminescence intensity at any instant after the cessation of excitation [68]. For the purpose of comparing different phosphors it would be very convenient to characterize the luminescence decay by one constant whose numerical value would give an explicit representation of the intensity variation at all stages of decay. However, of all the decay laws only the exponential law $I = I_0 e^{-T/T_0}$ satisfies this requirement. In this case the constant τ_0 gives both the time interval during which the luminescence decreases by a factor of e at any stage of decay (the decay rate is constant) and the mean lifetime of luminescence. Such a decay mode is encountered quite frequently in molecular systems. However, as has been shown, in crystal phosphors the process is much more complex. The determination of the mean lifetime of luminescence is frequently impracticable and it loses its practical value because this mean quantity contributes almost nothing to the determination of the actual luminescence intensity at any instant and it cannot be used to compare accurately the luminescence inertia of different phosphors. Consequently, instead of characterizing the luminescence inertia in terms of the decay law and the mean lifetime of luminescence, we have to use a time during which the luminescence intensity decreases by a certain factor relative to its original value.

This method for characterizing the inertia of the process answers to a certain extent the requirements of technology. It must be remembered, however, that different phosphors which decay by the same factor in a certain time interval may have very much different decay patterns before and after the particular fixed point of intensity reduction is reached.

The characteristic feature of complex recombination processes described by Becquerel's hyperbolic formula is an extremely fast decay of luminescence during the first stages of the process, followed by a slower decrease of intensity. Thus in the case of the decay law

$$I_t = 0.2 \ (0.2 + t)^{-1}$$

the luminescence intensity decreases from its original value by a factor of 10 in 1.8, by a factor of 100 in 19.8, and 1000 in 199.8 units of time in which the constant 0.2 is expressed. Thus each reduction of the luminescent brightness by one order of magnitude takes place in a time interval whose length also increases successively by a factor of about 10 [68]. In the general case the length of each decay step is $n^{(k-1)\alpha}(n-1)$, where n is the gradation of brightness reduction, k is its number, and α is the exponent in Becquerel's formula.

The slowly decaying "tails" of luminescence produced by successive excitation pulses become superimposed on each other and form a general background, which reduces the contrast between signals.

§ 2. Charge Localization Sites in the Lattice of Crystal Phosphors and Their Investigation

As has been shown above, the duration of the afterglow and its variation and temperature dependence are determined by the distribution of energy levels in which electrons and holes are localized in the lattice of the crystal phosphor. Thus the distribution of localized trapping levels is related to the main parameters characterizing the luminescent properties of the crystal phosphor.

The electron and hole trapping sites are created by crystal lattice imperfections in crystal phosphors. Their energy depth is determined by the nature of the host material of the crystal phosphors, by activators and coactivators introduced into it, and also by the temperature and duration of annealing. Since the localization sites of electrons and holes are not sufficiently numerous and do not exhibit any periodicity, they cannot be detected by x-ray analysis.

The highly-developed method of thermal excitation of phosphors [69, 70, 71] is most convenient for the study of the energy distribution and occupation of trapping levels. The phosphor is excited at a certain sufficiently low temperature and after the excitation is switched off the specimen is heated at a uniform rate. During the heating process electrons localized in deeper levels are raised to the conduction band. The thermoluminescence curve (t.l.c.), which shows the variation of the luminescence intensity as a function of uniformly increasing temperature, allows us to determine the existence of trapping levels of different energies, the nature of these groups of levels and their occupation by electrons under different conditions of excitation.

The energy level depths, corresponding to peaks, can be determined by a few methods [70, 72, 73, etc.] which, however, give appreciably different values. An approximate estimate of the level depth is given by a simplified formula of Randall and Wilkins [72]:

$$E = 25 \, kT,$$

where E is the level depth in electron volts and T is the absolute temperature. A peak on the t.l.c. is caused by a rapid, but not instantaneous, release of electrons from trapping levels emitting luminescent light. Consequently, phosphors emitting short-duration luminescence at room temperature should not exhibit any peaks on the t.l.c. at room and higher temperatures: the corresponding peaks appear at lower temperatures. The existence of peaks at room temperature leads to a long-duration afterglow. In general, peaks on the t.l.c. observed at a certain temperature correspond to levels which take part in the phosphor's luminescence at temperatures higher by 100° or more than the temperature of the peak [74].

The main advantages of the thermoluminescence method over other methods for investigating trapping centers are its simplicity and sensitivity. The thermoluminescence method is by a few orders of magnitude more sensitive than the incremental absorption method (this is evident from the well-known experimental fact that crystals in which incremental absorption cannot be observed exhibit an easily detectable afterglow and thermoluminescence).

It should be noted, however, that the initial distribution of electrons over the trapping levels, established during the excitation process in the phosphor, is given directly by the thermoluminescence method only in the case where the probability of recapture of electrons and holes released during heating is small in comparison with the recombination probability.

If repeated capture occurs frequently, then, knowing the electron transport direction (transition from shallow to deep levels during heating), from the ratio of the areas under the t.l.c. peaks in a number of cases we can draw certain semiquantitative conclusions concerning the initial electron distribution.

Using the thermoluminescence method, we can also obtain some information about the change in the electron distribution over the levels during the decay at a constant temperature. For this purpose a study is made of the t.l.c. (starting from the lowest temperatures) of an excited, rather than a decaying, phosphor and then of the t.l.c. of the same, though partially exhausted, phosphor whose decay is interrupted at different instants of the afterglow. The series of t.l.c. obtained in such a manner gives us information about the successive exhaustion of trapping levels and the partial transition of electrons from shallow to deep levels during the intrinsic decay process of phosphors [74, 75].

In the present work the thermoluminescence method was used to establish the relation between the luminescence inertia and the development of systems of trapping levels at different energy depths.

§3. Apparatus Used for Investigating the Luminescence Decay, Trapping Levels, and Other Properties of Crystal Phosphors during Electron Excitation

Apparatus for Studying the Cathodoluminescence Decay and the Energy Losses during the Electron Penetration through Thin Layers. To determine the decay curves of cathodoluminophors in a wide range of variation of the accelerating voltage and beam current density, and carry out the experiments described in Chapter II on the penetration and energy losses of electrons in zinc sulfide luminophors, we used an apparatus which is represented schematically in Fig. 38.

This apparatus is capable of producing an electron beam with energies of up to 50 keV and intensities between 10^{-10} and 10^{-6} A/cm^2. With the aid of a system of electrodes, forming the electron lenses 4, and the limiting diaphragms 6, the electron beam due to the thermionic emission from the tungsten cathode 2 is contracted into a narrow beam incident on the lower electron lens 8. This lens produces a magnified image of the diaphragm on the specimen 11 and we obtain thus a uniformly luminescent spot of known diameter.

The sample holder and Faraday cylinder (12, 10), mounted on porcelain insulators, are used for collecting all the secondary emission electrons, reflected electrons and the conduction current. This ensures accurate measurements of the beam current. The beam current is varied by changing the angular aperture of illumination and the filament current of the tungsten cathode.

The magnetic prism 5 is placed in the path of the electron beam. It corrects the direction of the illuminating system and is provided to prevent ion bombardment of the specimen. In the middle of the electron irradiating system the two plates 7 are mounted on porcelain insulators. They are used to deflect the electron beam away from the sample during measurements of the luminescence decay of the phosphor, and to provide pulsed excitation of the specimens investigated. Six specimens were introduced at once into the sample chamber and then they were investigated in succession.

The electron gun was powered with the high-voltage rectifier 14 consisting of kenotrons B-1$_{0.1/40}$ connected as a voltage doubler with a U-shaped RC filter. The high voltage was measured with an S-96 voltmeter

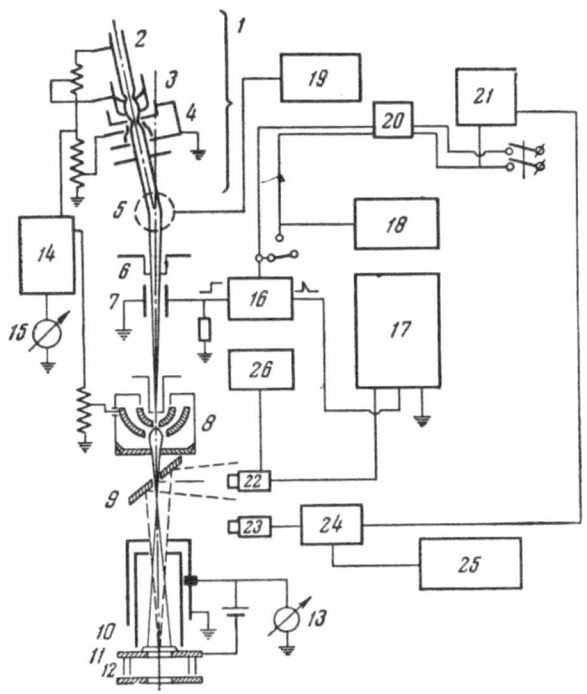

Fig. 38. Block diagram of the high-voltage apparatus for measuring the parameters of luminophors. 1) Illuminating system; 2) tungsten cathode; 3) modulator (Wehnelt cylinder); 4) electron lens; 5) magnetic prism; 6) diaphragm; 7) deflecting plates; 8) electron lens; 9) mirror; 10) Faraday cylinder; 11) sample; 12) carriage; 13) galvanometer; 14) high-voltage rectifier; 15) kilovoltmeter; 16) thyratron generator; 17) oscilloscope ÉNO-1; 18) stabilized rectifier VS-12; 19) magnetic prism power supply; 20) timing relay; 21) recording potentiometer ÉPP-09; 22 and 23) FÉU;24) paraphase amplifier; 25) amplifier power supply; 26) stabilized rectifier.

(15) up to 30 kV, and above 30 kV with the same meter using a 1:2 high-resistance voltage divider. The filament voltage was fed to the tungsten cathode from a stabilized source made from transistors. The filament power supply was placed in a well-insulated box, which contained all the components of the rectifier circuit operating at high voltages.

The beam was suppressed with a voltage supplied from the thyratron generator, which was triggered simultaneously with the sweep of the oscilloscope registering the luminophor's decay curve. The decay was measured by the method described below. Long-duration decay processes (longer than 10 sec) were measured with a circuit consisting of an automatic recording potentiometer (21, Fig. 38), which was switched in simultaneously with the generator suppressing the beam. Using this circuit, it was possible to cut off the excitation in 1×10^{-5} sec.

To supress the beam in 1×10^{-8} sec, the apparatus was provided with a high-voltage generator of square pulses with an amplitude of 1500 V and a front length of the order of 2×10^{-8} sec. Since at 50 kV the pulse amplitude required for a complete deflection of the beam was smaller than that of the pulse produced by the generator, the blocking length was much shorter than the rise of the pulse front. The pulse generator provided single or repeated pulses with a variable off-duty factor. The pulse length was variable between 0.5 and 10 msec. The generator described can be used for measuring very short initial stages of decay of cathodoluminophors.

The apparatus was designed to measure the parameters of luminophors for specimen temperatures between -196 and $+200°C$. For this purpose the specimen holder was replaced with a special attachment made from industrial copper, and the specimen under investigation was placed inside it. The attachment was insulated from the apparatus housing, so that it was possible to measure the excitation intensity of the phosphor. The temperature of the phosphor was varied using a copper heat conductor, connected to the attachment for holding the specimen. For cooling, the heat conductor was placed in a vessel with a liquid coolant, and for heating, an electric heater mounted on a quartz tube was inserted on the conductor. The apparatus was suitable for reflection measurements on specimens in the form of powder and also for transmission and reflection measurements on screens.

Apparatus for Measuring the Decay, Spectra, and Brightness of Cathodoluminescence. An apparatus suitable for measuring ten specimens during one cycle was used for carrying out a series of determinations of the decay curves, spectra, and brightness of crystal phosphors. The stimulus was supplied by a fixed defocused electron beam incident on the screen at an angle of 45° and forming an elliptic luminous spot on it. The filament of the electron irradiating system was powered with a high-frequency generator, producing a voltage with a frequency of 50 kc. The use of the high-frequency generator eliminated decay curve distortions caused by the pulsation of the filament current, whose period in the case of the 50 kc filament voltage was close to the duration of the decay in rapidly de-excited phosphors.

The accelerating voltage was applied between the cathode and the specimen chamber, which acted as the anode of the apparatus. The specimen chamber was insulated from ground for the entire accelerating voltage. The cathode of the system was at the "ground" potential. Using this circuit, it was easy to suppress the beam by applying to the first focusing electrode of the irradiating system a negative potential relative to the grounded cathode. A thyratron relay with an artificial field (Fig. 39) was used for suppressing the beam. The focusing electrode of the irradiating system was connected to the slide of the potentiometer R_3. When the circuit between points A and B is broken, the voltage on the focusing electrode relative to the ground may be varied with the potentiometer R_3 from -25 to $+15$ V, until the best focusing of the beam is achieved.

The key K closes the anode circuit of the thyratron. A positive pulse lagging behind the oscilloscope sweep is applied from an ÉNO-1 oscilloscope to the second grid of the thyratron. After the key K is closed, the first pulse triggers the thyratron and the voltage between the points A and B reduces to the voltage drop across the thyratron. The voltage at the point B and on the focusing electrode becomes 210 V relative to the "ground."

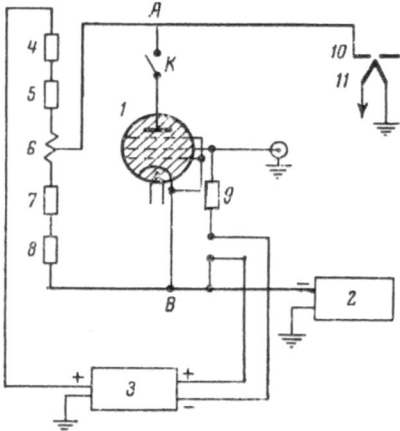

Fig. 39. Beam suppressor circuit for measurements of the afterglow.
1) Thyratron TGZ $-01/1.3$; 2) stabilized rectifier VS-11; 3) stabilized rectifier VS-12; 4, 5, 6, 7, 8, 9) resistors: $R_1 = 560\ \Omega$; $R_2 = 4\ k\Omega$; $R_3 = 21\ k\Omega$; $R_4 = 20\ k\Omega$; $R_5 = 560\ \Omega$; $R = 100\ k\Omega$; 10) modulator; 11) tungsten cathode.

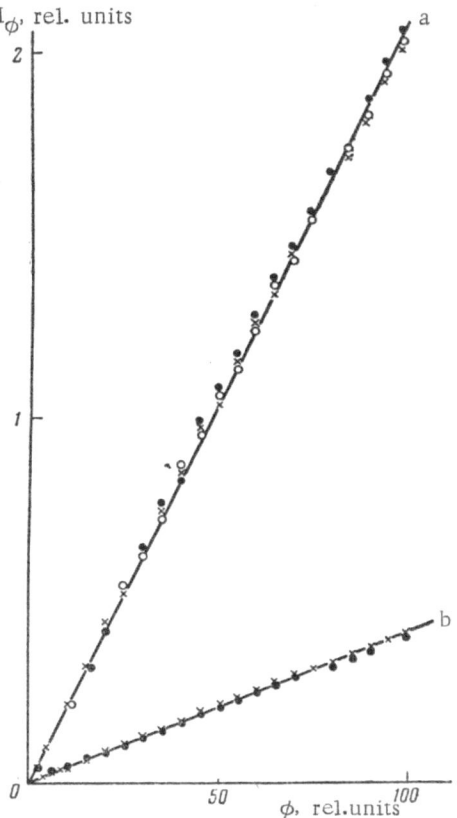

Fig. 40. Luminance characteristics of the FÉU-14B obtained for voltages corresponding to different sensitivities of the FÉU. a) 1:100; 1:10; 1:1; b) 1:50; 1:5. One unit of the scale along the axis of ordinates for the 1:100 sensitivity corresponds to 10 relative (rel.) units; for the 1:5 and 1:10 sensitivities it corresponds to 100 rel. units; for the 1:1 sensitivity it corresponds to 1000 rel. units.

The beam is completely suppressed even by -50 V on the focusing electrode. The slope of the suppressing pulse front is determined by the warm-up time of the thyratron and an RC circuit consisting of resistors R_3, R_4, and R_5, and the connecting cable. The transient capacitance of the thyratron is negligible. All the connections were made with an RK-50 cable, which has a capacitance, according to specifications, no greater than 27 pF per meter. Measurements of the rise rate of the suppressing pulse directly at the output contact of the Wehnelt cylinder showed that the rise time of the front of the suppressing pulse is much shorter than 5×10^{-5} sec.

The beam current was estimated from the secondary electron current incident on the specimen chamber. Since the resistivity of the overwhelming majority of luminophors is very high, during the bombardment of a "thick" luminophor layer (thickness about 1 mm) with an electron beam the main agency by means of which electrons can leave the phosphor is secondary emission. By secondary electrons we understand not only the secondary emission electrons emitted from the bombarded surface, but also the electrons reflected elastically and inelastically. The secondary emission coefficient of luminophors depends on the electron energy. If the secondary emission coefficient is different from unity, charges accumulate on the surface until its potential reaches a value for which the secondary emission coefficient is unity. Thus the secondary electron current incident on the specimen chamber is almost the same as the beam current.

To measure the beam current, we used a vacuum tube electrometer representing a two-cascade dc amplifier. The voltage was measured with a microammeter calibrated with an S-96 electrostatic kilovoltmeter. The microammeter was connected into the circuit of the high-resistance divider. The measurements of the luminescence characteristics of phosphors were made at current densities between 1×10^{-9} and 1×10^{-6} A/cm^2 and voltages between 5 and 20 kV. The apparatus was equipped with a metal vacuum system producing a vacuum of the order of 10^{-5} mm Hg in the working part of the specimen chamber.

A cathode-ray oscilloscope ÉNO-1 was used to determine the decay curves. The radiation from the luminophor passed through the quartz window in the specimen chamber and an optical attachment and was received by the cathode of a photomultiplier. The photocurrent without any additional amplification was fed through a short length of the RK-50 cable to the input of the ÉNO-1 oscilloscope. When the sweep frequency was selected properly, decay curves were clearly visible on the oscilloscope screen at the instant when the beam was suppressed. However, it was possible to use one curve to study only the decay of up to a few percent of the original brightness.

To investigate the decay in a wider range of variations of the brightness intensity, parts of the decay curve were observed on different scales. The scale was changed by varying the voltage on the photomultiplier. The voltage on the FÉU was supplied through a switch. Connected to the four positions of the switch were variable resistors, selected in such a way that upon switching from one resistor to another the magnitude of the signal was changed by a factor of 5, 10, 50, or 100.

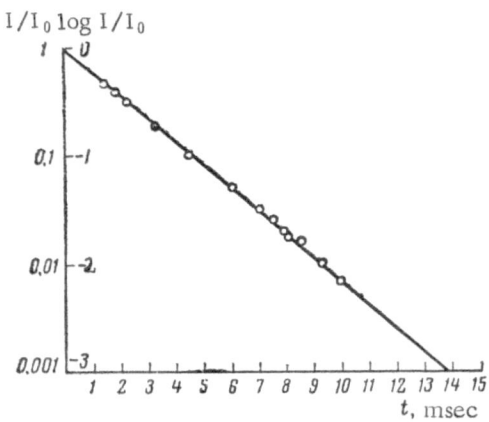

Fig. 41. The luminescence decay curve of a ZnS-Mn phosphor on a semilog scale.

The entire decay curve was located on the oscilloscope screen only for the sensitivity of 1 : 100, but in this case the end stages of decay were not visible. At higher sensitivities of the FÉU, at the instant when the luminophor was excited the luminous spot of the oscilloscope was outside the screen. At the instant when the beam was suppressed the initial stages of decay were cut off. The screen showed only a part of the decay curve; the higher the sensitivity of the FÉU, the smaller the visible part.

Curves obtained at five different sensitivities of the FÉU were photographed and then transposed with the aid of a photomagnifier onto a graph paper. The luminescence decay curve of the luminophor was constructed from the curves obtained by taking into account the intensity and time scales of the image. Since the curves of the decay as a function of time obtained in this manner coincided with each other, it was possible to check whether the selection of the FÉU voltage was proper or not.

The linearity of the luminance characteristics of the photomultiplier at all operating voltages is very important from the point of view of proper transferring of the shape of the decay curve. The luminance characteristics of the FÉU-14B photomultiplier are shown in Fig. 40. The luminous flux in relative units is plotted along the axis of abscissas and the photocurrent in relative units along the axis of ordinates.

The vertical scale of the ÉNO-1 oscilloscope was checked for linearity. According to the specifications, the instrument is capable of measuring the pulse amplitude to an accuracy of $\pm 10\%$. This accuracy was not sufficient. Using an M-91 microammeter we constructed a calibration curve for correcting the decay curves for the nonlinearity.

The time scale was checked with a generator of sinusoidal oscillations and thus it was possible to measure frequency to an accuracy of $\pm 1\%$. The uniformity of the frequency characteristic of the oscilloscope was also checked. The vertical deflection amplifier of the oscilloscope had a pass band extending from 0 to 10^{-6} cps.

The use of the above-described method for sensitivities 1:50, 1:10, 1:5, and 1:1 resulted in overloading of the oscilloscope during the initial stages of decay. However, it was established by special experiments that even an overload by a factor of 100 did not result in an appreciable distortion of the pulse.

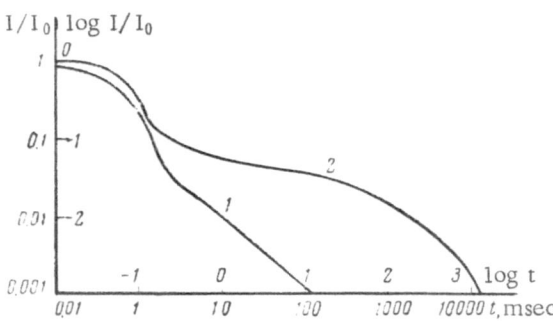

Fig. 42. Luminescence decay curves of the ZnS-Tu phosphor obtained with different types of photomultipliers. 1) FÉU-42; 2) FÉU-24.

To check the accuracy of the decay curves, we obtained a decay curve for the ZnS-Mn phosphor, which is known to decay exponentially during the initial stages [76]. We obtained an almost perfect straight line on the semilog scale (Fig. 41).

During our selection of the proper type of a photomultiplier for the afterglow measurements, we observed that certain types of photomultipliers have a large time constant. We were concerned about the true reproduction of the rapidly varying process, rather than the time resolution of individual flashes.

Figure 42 shows two decay curves for the same phosphor obtained with FÉU-42 and FÉU-24 types of photomultiplier. Since both these photomultipliers have antimony-cesium cathodes, the difference between the curves cannot be related to different spectral sensitivities of the photocathodes. However, the time required for the initial brightness to decay to 0.001 of its value, as measured with these two FÉU, differed by a factor of more than 100.

Whereas the decay curves obtained with the FÉU-42 were reproducible, the operation of the FÉU-24 was unstable. In addition to the FÉU-24, the FÉU-19, FÉU-34, and FÉU-49 also exhibit a time lag. The shape of decay curves was well reproduced with the FÉU-33, FÉU-14A, FÉU-14B, FÉU-42, and FÉU-1B. Even though the reason for the photomultipliers' time lag was not investigated specially, from a comparison of the construction of the photomultipliers we concluded that this effect is caused by insufficient shielding of the FÉU dynodes from glass and other insulators.

The spread in the time required for the phosphor to decay to 0.1% of its intensity at the instant of excitation, as determined from repeated measurements by the method described, was $\pm 15\%$. This method allowed us to obtain the true luminescence decay curve for intensities differing by a factor of 1000 from the luminescence level at the instant of excitation. The time interval for measurements was determined by the selection of the sweep velocity of the oscilloscope ÉNO-1 and was variable from 10 μsec to 10 sec.

Apparatus for Investigating the Thermal De-excitation and for Measuring the Luminescence Time at Different Temperatures. The thermoluminescence curves were obtained with the apparatus shown in Fig. 43. The apparatus was constructed in such a way that it was possible to investigate also the brightness and the inertial characteristics of luminophors in a wide range of accelerating voltages and beam current densities at temperatures from -196 to $+300°$C.

The metal Dewar vessel 8 was used as a chamber for the specimens under investigation. The massive copper cube 9 was firmly attached at the bottom of the Dewar vessel cylinder. It provided the required thermal conditions for the luminophor. The specimen was cooled by filling the Dewar vessel with liquid nitrogen and a special electric heater was used for heating.

The temperature was measured with a calibrated thermocouple attached to the copper cube.

This construction of the apparatus allowed us to do the following: 1) change specimens without disturbing the fixed position of the phosphor under investigation; 2) excite the phosphor at any temperature from -196 to $+300°$C and maintain this temperature to an accuracy of $\pm 2\%$ for 10 to 15 min; 3) achieve a very uniform rate of heating the phosphor during the test; 4) excite the specimen either with a cathode or a light beam (mercury lamp with filters passing the 366 or 313 mμ line).

The electronic circuit of the apparatus is shown in Fig. 44. An irradiating system consisting of the filamentary cathode 1, the focusing electrode 2, and the first anode 3 was used to obtain the electron beam.

The luminophors' radiation was collected with an FÉU-14B, whose signal was fed through a paraphase amplifier to a multipoint potentiometer ÉPP-09, recording automatically both the t.l.c. and the temperature variation, or to a microammeter M-95.

In order to determine the distribution of trapping levels, the specimens were excited for 20 min so that the t.l.c. could be obtained. At a current density $j = 3 \times 10^{-8}$ A/cm^2 the trapping levels became completely saturated. The thermoluminescence measurements were made 15 min after the cessation of the excitation. The heating rate was 10 deg/min. This rate gave a good resolution of individual peaks on the t.l.c. and a sufficient brightness of the phosphor's thermoluminescence.

The same apparatus was used to measure the decay of the luminophor specimens at different temperatures.

Since the cathode and the modulator in the apparatus described were at a high potential, for suppressing the beam we used a specially designed system of deflecting plates (5, Fig. 43), to which the deflecting pulse from the thyratron circuit was applied. The electron-optical system of the apparatus was designed in such a way that the electron beam leaving the irradiating system (1, Fig. 43) was focused with an additional magnetic coil (2, Fig. 43) and then deflected through an angle of 30° from its original direction (because the beam axis was rotated through 30° relative to the plane of the luminophor specimen). Because of the deflection of the beam, it was possible to adjust the electron beam accurately in relation to the specimen under investigation and observe the luminescence of the specimen. Also, the deflection of the beam eliminated completely

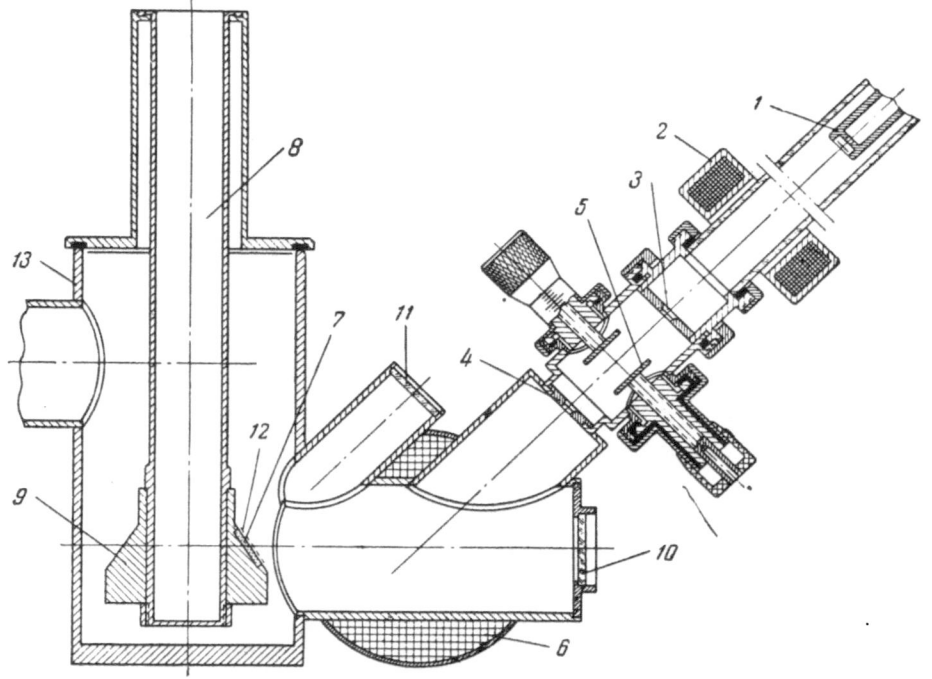

Fig. 43. Diagram of the high-voltage apparatus for investigating the trapping levels and decay of luminophors at different temperatures. 1) First anode of the electron gun; 2) magnetic focusing coil; 3, 4) diaphragms; 5) deflecting plates for suppressing the electron beam; 6) rotating coil; 7) specimen; 8) Dewar vessel; 9) copper cube; 10) observation window; 11) photoexcitation window; 12) grid for current measurements; 13) housing.

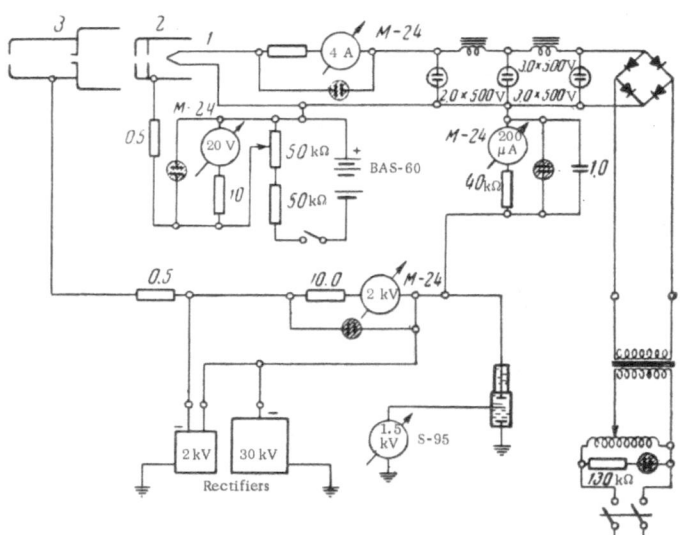

Fig. 44. Electronic circuit of the apparatus for investigating the trapping levels and decay of luminophors.

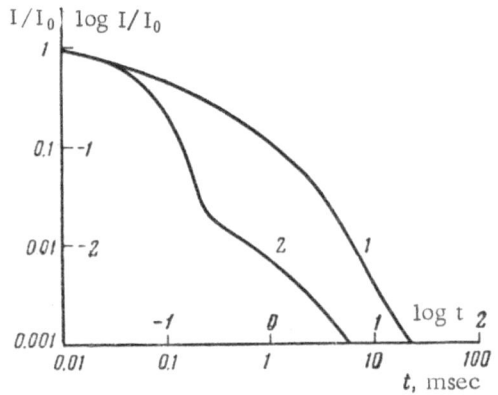

Fig. 45. Simple (1) and complex (2) decay curves.

the effect of the ionic bombardment of the luminophor.

The electron beam field in front of and behind the deflecting plates was limited by removable diaphragms (4, Fig. 43), which together with the magnetic focusing coil produced the necessary diameter of the electron beam in the space between the deflecting plates.

The deflecting plates were constructed in such a way that it was possible to change their separation and their mutual position relative to the electron beam axis while the electron gun was operating, without disturbing the vacuum. Such a construction of the deflecting system for suppressing the electron beam lowered the requirement of accurate adjustment of the electron beam and ensured a high rate of suppressing.

The VS-11 and UIP-1 connected in series were used as the power supply for the thyratron circuit. They ensured a flow of high current (about 50 mA) through the thyratron at voltages up to 900 V.

Using the suppressing circuit described, it was possible to cut off the excitation of the luminophors very quickly (in about 1×10^{-6} sec) even for large spacings between the plates.

The trigger pulse was fed to the thyratron generator from an ÉNO-1 oscilloscope, while the measured signal from the FÉU-14B was fed to the input of the oscilloscope.

§4. Dependence of the Luminescence Decay Mode of Phosphors on the Distribution of Trapping Levels

To establish directly the effect of individual trapping levels on the natural decay mode of phosphors and on the duration of their afterglow, we investigated the decay and t.l.c. of ZnS-Ag, ZnS-Tu, phosphates, and other luminophors.

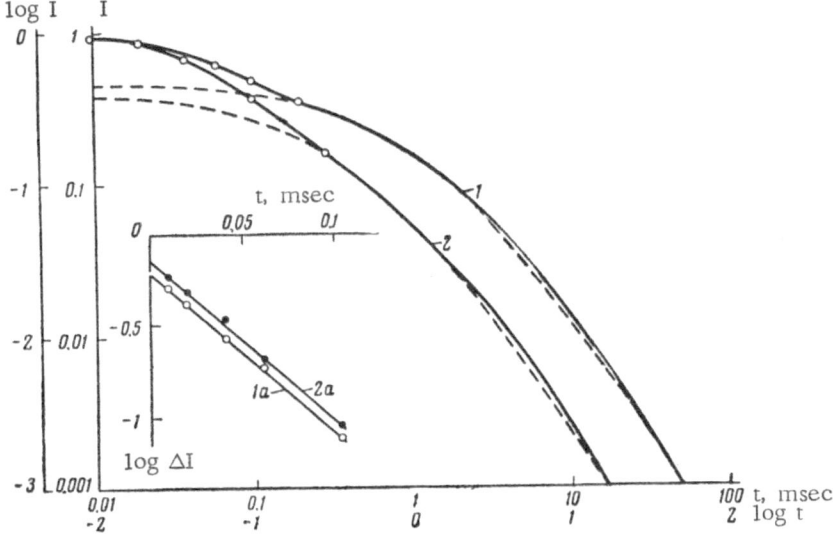

Fig. 46. Decay curves of the ZnS·CdS-Ag phosphor for different excitation current densities: 1) 10^{-7} A/cm^2; 2) 5×10^{-6} A/cm^2. The solid lines represent the experimental curves and the dashed lines the Becquerel curves. The points at the beginning of the solid curves correspond to the sum of the calculated exponentials and Becquerel curves.

TABLE 3

j, A/cm^2	a	b	α	I_0	$\tau_0 \cdot 10^{-5}$, sec
10^{-7}	1.672	1.667	1.545	0.60	4.8
$5 \cdot 10^{-7}$	1.736	2.642	1.54	0.57	4.8
10^{-6}	1.760	2.985	1.57	0.59	5.2
$5 \cdot 10^{-6}$	1.865	4.818	1.56	0.71	4.6

As far as the shape is concerned, the decay curves of the phosphors investigated can be divided into two groups — simple and complex (Fig. 45). In the log-log coordinates the simple decay curves are monotonic. They are characterized by a slow decrease of the brightness during the first stage of decay and a rapid one during the later stages; this second part of the curve can be replaced with a straight line intersecting the axis of abscissas at a certain angle α (about 45 to 50°); the magnitude of $\tan \alpha$ characterizes the decay rate during the second stage. The simple curves can be approximately represented by E. Becquerel's formula

$$I = A\,(b + t)^{-\alpha}.$$

The shape of the complex curves is shown by curve 2 in Fig. 45. In this case a stronger, more rapidly decaying process is superimposed on the described monotonic process during the initial stages of decay. Consequently, a kink appears on the decay curve. Such a complex luminescence decay mode is well-defined in phosphors activated with Tu and in phosphates. However, a more detailed analysis shows that even the smooth curves deviated appreciably from the hyperbolic form.

We will analyze in more detail the case of the complex decay curves. Both for the ZnS-Tu phosphors and for phosphate luminophors activated with Eu the decay curve (in the log-log coordinates) splits into two segments. The first segment, where the luminescence decay is more rapid, terminates for ZnS-Tu at about 0.15 msec and for phosphates at about 0.18 msec. In this interval the first, more rapid, process plays the main role. During the later stages of decay the main luminescence is caused by the second process. The resolution of the decay curve of the ZnS-Tu phosphor into the short and long-duration processes shows that the long-duration process is accurately described by the hyperbolic formula $A(b + t)^{-\alpha}$, while the short-duration process may also be expressed by a hyperbolic formula whose constants are, however, much different from the constants of the first process. In certain phosphors the initial decay process, separated out of the curve, obeys an exponential law.

Figure 46 shows a log-log plot of the decay curves of the ZnS·CdS-Ag phosphor for different excitation current densities (curve 1, 10^{-7} A/cm^2; curve 2, 5×10^{-6} A/cm^2). The dashed lines represent the Becquerel curves $I = A(b + t)^{-\alpha}$, whose constants were determined from experimental points. The logarithms of the differences between the ordinates of the actual experimental curves and of the corresponding values of intensity given by the Becquerel curves were plotted on the semilogarithmic scale (curves 1a and 2a). It is evident from the figure that these differences may be described quite accurately by exponentials.

A similar analysis of the decay curves was made also for current densities 5×10^{-7} and 1×10^{-6} A/cm^2. In all cases it was possible to represent the decay curve by the sum of a hyperbola and an exponential:

$$I = (a + bt)^{-\alpha} + I_0 e^{-\frac{t}{\tau_0}}.$$

The values of the constants appearing in this formula are shown in Table 3.

The values of the luminescence intensity used in our calculations were normalized. It follows from the values of the constants that the decay of the short-duration exponential process is independent of the current density. The rate of the hyperbolic process during the

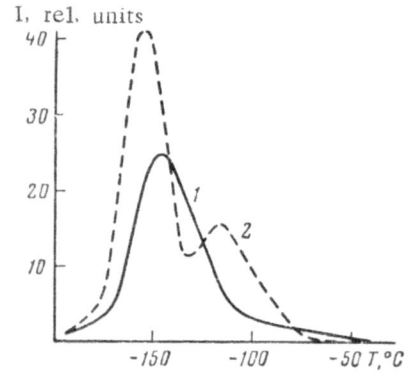

Fig. 47. The t.l. curves of luminophors. 1) ZnS·CdS-Ag; 2) ZnS-Tu.

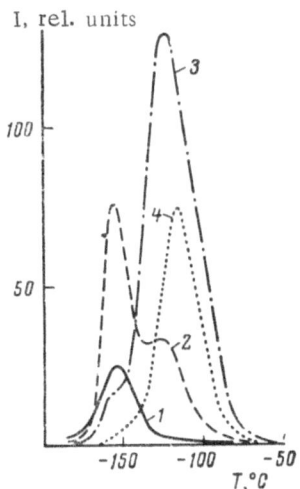

I, rel. units

Fig. 48. The t.l. curves of the ZnS-Tu, Ni luminophors with different contents of Ni. 1) 0%; 2) 10^{-7} g/g; 3) 10^{-6} g/g; 4) 10^{-5} g/g.

first stages increases with increasing current density (b increases); during the later stages the decay is the same in both cases.

It must be noted that in phosphors with two luminescence processes a reduction of the integral luminescence by a factor of n corresponds to a reduction of the brightness of the long-duration process by a factor of $n/(1+k)$, where k is the ratio of the initial intensity of the short-duration process to the initial intensity of the long-duration luminescence. The larger the value of k, the smaller the factor by which the intensity of the long-duration process is reduced during the time required for the integral luminescence to reduce by a factor of n. During the later stages of decay, on the other hand, only the long-duration process is important; consequently, the time required to reduce the integral luminescence by a factor of n also becomes much shorter.

To establish the relation between the shape of the decay curves and the position of the trapping levels of phosphors, we obtained their t.l.c. (Fig. 47). From the comparison of the decay curves with the t.l.c. of the ZnS-Tu phosphors we can conclude that the complex shape of the decay curves is related to the presence of two peaks on the t.l.c. Two stages of the afterglow may appear on the decay curves as a result of different rates of electron release from two different systems of levels; the release of electrons from shallow levels and their recombination are satisfactorily described by an exponential function, and this may be related to the fact that recombina- is more probable than recapture.

For the ZnS-Tu phosphors and phosphates activated with the rare earth elements the short-duration process may also be given a different interpretation, by attributing it to the excitation of the luminescent centers. However, this idea meets with the objection that the decay curves of these phosphors cannot be readily separated into the exponential and hyperbolic parts. The effect of the position of the trapping levels on the shape of the decay curves is quite evident from the comparison of the phosphate and ZnS-Tu phosphors. The depth of the first trapping level both in ZnS-Tu and in strontium phosphate has the same value of 0.25 eV, but the second

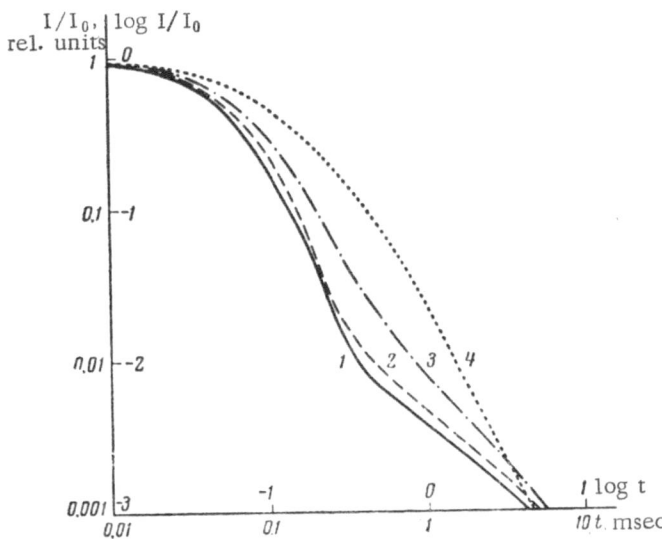

Fig. 49. Decay curves of the ZnS-Tu, Ni luminophors with different contents of Ni. 1) 0%; 2) 10^{-7} g/g; 3) 10^{-6} g/g; 4) 10^{-5} g/g.

TABLE 4

Ni con-centration, g/g	Decay time, msec		
	to 0.1	to 0.01	to 0.001
—	0.16	0.37	4.2
10^{-7}	0.17	0.47	5.2
10^{-6}	0.23	0.86	5.6
10^{-5}	0.40	1.5	4.7

level in ZnS-Tu is at the depth of 0.34 eV, while in strontium phosphate it is at 0.4 eV. The fact that the second level in the phosphate luminophor is deeper explains the later occurrence of the inflection point on the decay curve.

While comparing the decay curves with the t.l.c. levels we must take into account the fact that on the graphs presented the first ones are normalized and the others are given in their natural form. To make a proper comparison of the effect of the levels with the luminescence curves, we should normalize also the latter. All slowly decaying phosphors have a well-developed system of deep levels which are active during the later stages. At higher excitation densities the shorter-duration process is more important and the decay curve smooths out.

In silver-activated phosphors there is only one level and this leads to a smooth decay of the luminophors (curve 1, Fig. 45). The successive change of the system of trapping levels in the case of the series of ZnS-Tu, Ni phosphors with different contents of Ni allowed us to compare the variation of the t.l.c. with the variation of the decay mode.

As the percentage content of Ni is increased, the t.l.c. changes its shape. Instead of the two peaks appearing in the absence of Ni and at low concentrations of Ni equal to 10^{-7} g/g (Fig. 48), only one peak is evident. At the same time, the shape of the decay curve also changes (Fig. 49), as it becomes smoother. At a nickel content of 10^{-5} g/g (where one peak appears on the t.l.c.) the decay curve becomes simple.

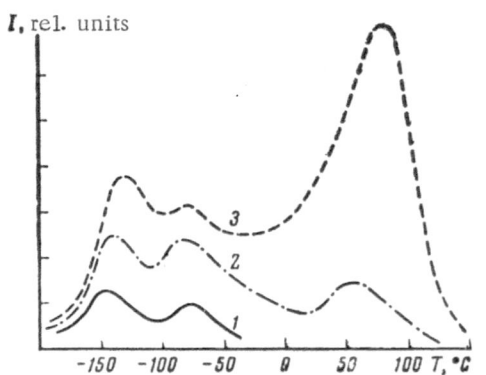

I, rel. units

Fig. 50. T.1. curves of phosphates. 1) $Sr_3(PO_4)_2$-Eu; 2) $Sr_3(PO_4)_2$-Eu, Gd; 3) $Sr_3(PO_4)_2$-Eu, Ce.

Data on the time during which the brightness of these phosphors decays to 0.1, 0.01, and 0.001 of the original value are shown in Table 4.

Moreover, as the nickel content is increased, the center of gravity of the t.l.c. shifts in the direction of high temperatures and this results in a change of the time required for the brightness to decay from its value at the instant of excitation to a certain value. As is evident from Table 4 these changes are not the same for the decay to 0.1, 0.01, and 0.001 of the original brightness.

The increase in the decay time of these phosphors during the initial stages of brightness reduction (to 0.1 and 0.01 of the original brightness) is due to the reduced importance of shallow levels, which vanish upon the introduction of nickel.

The relation between the t.l.c. and the decay mode of the phosphate phosphors is especially well-defined. The introduction into the phosphate phosphors of additional rare earth elements, Gd and Ce, results in the appearance of new deep trapping levels (Fig. 50), which bring about an increase in the duration of the afterglow and a change in the shape of the decay curves, manifested in new points of inflection (Fig. 51).

Finally, it should be noted that during the very initial stages of decay a region of rapid decrease of the luminescence, 2.5×10^{-5} sec long, can be observed in all the types of luminophors investigated at different temperatures and excitation densities.

We can assume that in this region recombination of electrons from the shallowest levels (not registered on our thermoluminescence curves) and of electrons from the conduction band takes place.

§5. Dependence of the Stored Light Sum and Decay on the Excitation Density

It is known that the rate of phosphorescence decay in almost all cathodoluminophors increases as the excitation density of the cathode beam is increased [4, 3]. In the case of the photoluminescence the increase of

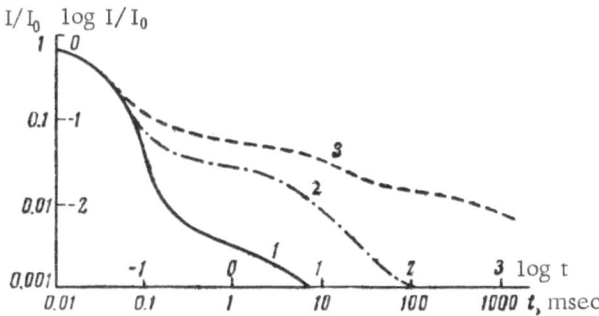

Fig. 51. Decay curves of phosphates. 1) $Sr_3(PO_4)_2$-Eu; 2) $Sr_3(PO_4)_2$-Eu, Gd;
3) $Sr_3(PO_4)_2$-Eu, Ce.

the decay rate with increasing excitation intensity is explained by the fact that, as the excitation density increases, the shallow levels become occupied to a greater extent and, consequently, their importance increases during the first stages of decay. The same effect is also present in the case of cathode excitation, but in the latter case the luminescence stimulation of the electron beam exerts a large additional influence. Owing to the electrical nature of the interaction between the excitation and localized electrons, the release of the localized electrons is faster in this case than when light quanta provide the stimulation [77]. This luminescent effect influences greatly the filling of deep trapping levels, because as far as the electrons in shallow levels are concerned, it must compete with the thermal excitation of electrons into the conduction band.

To understand the observed processes we must state here that, in addition to levels which we investigated at above −180°C, in many phosphors, in particular in phosphors of the ZnS-type, there are also shallower levels.

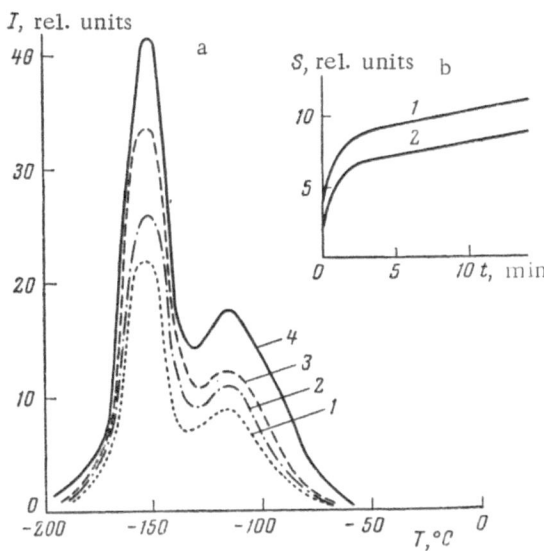

Fig. 52. The t.1. curves (a) and the variation of the light sum (b) of the ZnS-Tu luminophor as functions of the excitation time. a: V = 15 kV, j = 3 x 10^{-8} A/cm^2.
1) 3 sec; 2) 7 sec; 3) 1 min; 4) 15 min. b: 1) Increase of S in the 0.25 eV level;
2) increase of S in the 0.34 eV level.

111

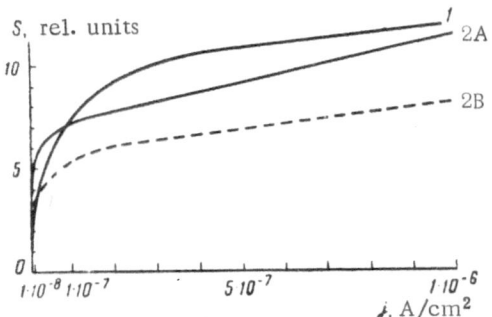

Fig. 53. Dependence of the light sum stored in trapping levels on the current density. 1) ZnS·CdS-Ag; 2A) the 0.25 eV level of ZnS-Tu; 2B) the 0.34 eV level of ZnS-Tu.

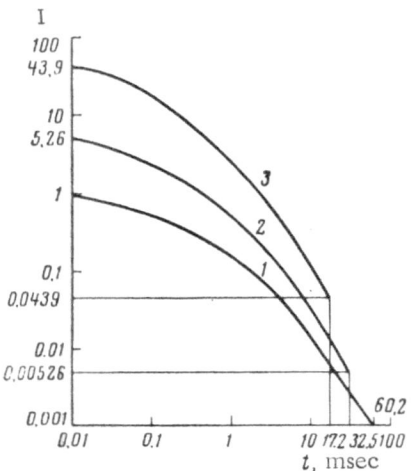

Fig. 54. Decay curves of the ZnS·CdS-Ag luminophor at different current densities. 1) 1×10^{-7} A/cm²; 2) 5×10^{-7} A/cm²; 3) 5×10^{-6} A/cm².

To check the hypothetical reasons for the increase of the decay rate, we compared the degree of occupation of the trapping levels with the nature of the decay for different excitation densities. Such an investigation is of interest, because practically no work has been done on the comparison of the luminescence inertia with the light sum stored during excitation.

The light sum stored in trapping levels during short-duration excitation processes should depend on the excitation time. This has been confirmed experimentally in a number of papers, including [71, 78]. However, after a certain excitation time is exceeded we reach a limiting light sum which remains almost constant as the excitation is further increased.

The results of our measurements of the light sum at different excitation densities showed that the same limiting light sum is obtained for both weak and strong excitation in cooled phosphors.

The time during which the light sum reaches saturation decreases as the excitation density increases. These results show that, in order to study the rate of occupation of trapping levels as a function of the excitation density, the luminophor must be excited in a very short time. During the study of the effect of the current density on the occupation of levels, we selected an excitation time of 30 sec. Our measurements showed that an excitation of this duration clearly reveals the nature of the occupation of levels over a wide range of current densities from 10^{-9} to 10^{-6} A/cm². In addition, this time interval approximately corresponded to the excitation time used to obtain the afterglow curves of the phosphor. The latter circumstance becomes important during the comparison of the nature of the decay with the occupation of trapping levels.

Figure 52 shows the t.l.c. for the ZnS-Tu luminophor obtained for different electron beam excitation times (V = 15 kV, j = 3×10^{-8} A/cm²) and the variation of the light sum as a function of the excitation time, determined from the area under the corresponding t.l.c. peaks. The figure shows that the distribution of localized levels of the phosphor has two peaks with maxima at −155 and −115°C (the energy depths are 0.25 and 0.34 eV respectively).

Upon excitation both trapping levels become occupied at the same rate. The height of the peak at −115°C always remains lower by a factor of 2.5 ± 0.1 than the height of the peak at −155°C, but the halfwidth of the band at −155°C is 0.53 of the band at −115°C. Consequently, the ratio of the stored light sum is 1.34.

When S is represented as a function of log t, the experimental points form straight lines intersecting the log t axis at one point, whence in the investigated range of excitation times S may be described by the formula

$$S = a_i \log bt,$$

where a_i is the slope of the straight lines and b is a quantity which is constant for both types of levels. In the case of our two levels i is 1 and 2 and $a_1/a_2 = 1.34$. This interpolation formula is not valid, of course, for the initial stages of storing the light sum.

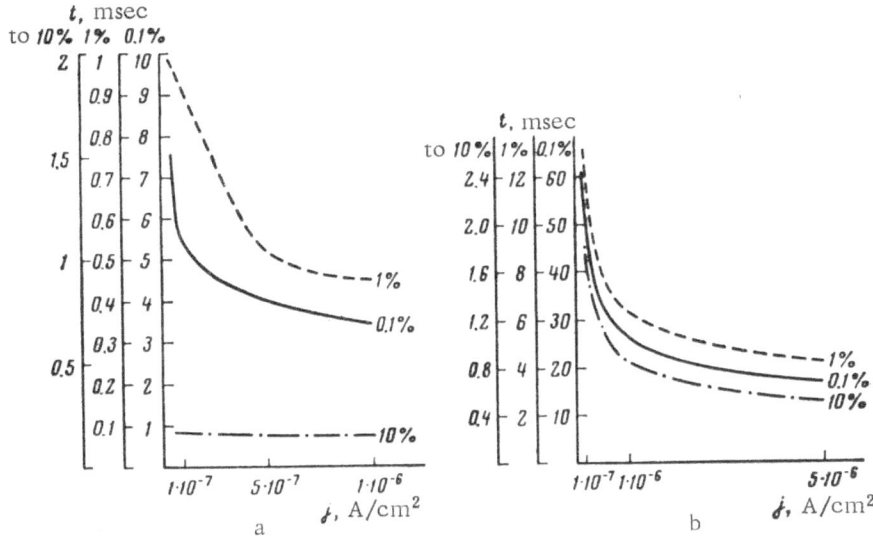

Fig. 55. Dependence of the decay time of the ZnS-Tu (a) and ZnS·CdS-Ag (b) luminophors on the current density.

The ratio of the light sums of the two levels, equal to about 1.3, remains the same for the ZnS-Tu luminophor also in the case when they increase with increasing current density of the electron beam. The highest rate of increase of the light sum is in the range of current densities between 10^{-9} and 10^{-7} A/cm². At j = 10^{-8} A/cm² the light sums stored in the trapping levels approach the limiting value. It should be noted that in all the luminophors the variation of the light sum with the excitation density, starting from j = 10^{-7} A/cm², is very slight (Fig. 53), because at these densities the number of electrons that can accumulate in these levels in 30 sec is close to the limiting value. As the excitation time is increased, the light sums remain almost constant.

And yet the original afterglows of luminophors excited with electron beams are very different. This indicates that shallower levels take part in the decay of the systems.

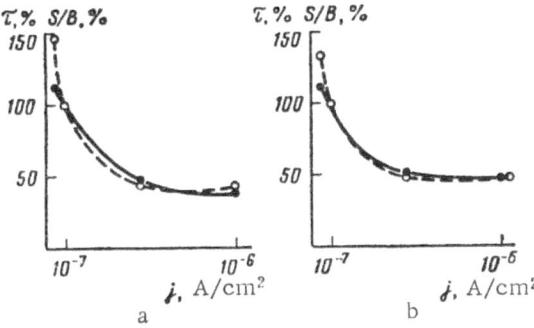

Fig. 56. Relative change of $\tau_{0.001}$ (solid curve) and of the quantity S/B (dashed curve) as a function of the current density (the values of τ and S/B at j = 1×10^{-7} A/cm² are taken as 100%). a,b) Specimens of the ZnS·CdS-Ag and ZnS-Ag luminophors.

The decay curves of a number of specimens and the reduction in the time required for the decay to certain values (to 0.1, 0.01, and 0.001) of the original brightness are shown in Figs. 54 and 55.

As a rule, the decay time of all the types of luminophors decreases as the excitation density increases. The exception is the ZnS-Tu phosphor in which the decay time to 0.1 of the original brightness is almost independent of the current density. As has been shown above, a special short-duration process develops during this stage of decay. To this process we must ascribe the property of being independent of the electron beam density. The decay curves of the ZnS·CdS-Ag luminophor for current densities 1×10^{-7}, 5×10^{-7}, and 5×10^{-6} A/cm² are shown on the log-log scale in Fig. 54. The horizontal lines correspond to the 0.001 level of the original brightness for each excitation density. In this range of current densities the decay time decreases from 60.2 to 17.2 msec.

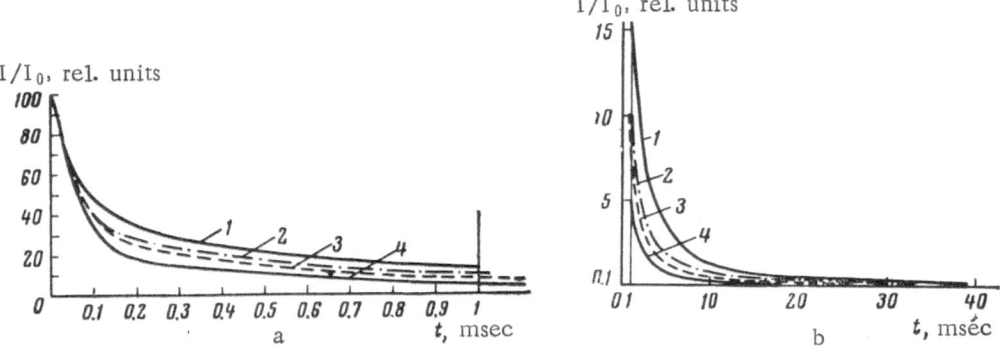

Fig. 57. The initial part of the decay curves of the ZnS·CdS-Ag luminophor in the normal coordinates for different current densities (a) and a later stage of decay — after 1 msec (b). 1) $j = 10^{-7}$ A/cm^2; 2) $j = 5 \times 10^{-7}$ A/cm^2; 3) $j = 10^{-6}$ A/cm^2; 4) $j = 5 \times 10^{-6}$ A/cm^2.

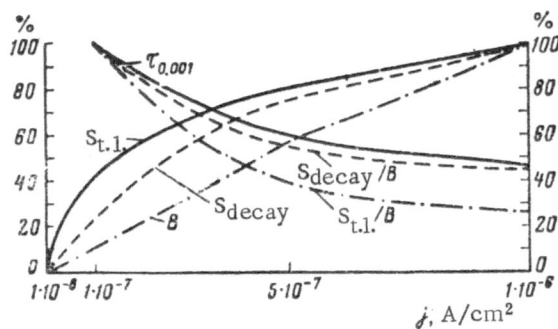

Fig. 58. The relative change of the original brightness (B), the decay time to 0.001 of the original brightness ($\tau_{0.001}$), the light sum from the t.l.c. ($S_{t.l.}$), the light sum from the decay (S_{decay}), and the quantities $S_{t.l.}/B$ and S_{decay}/B for the ZnS·CdS-Ag luminophor at different current densities.

To examine the reasons for the increase of the decay rate of the luminophor specimens with increasing density of the exciting electron beam, we compared the variation of the steady state brightness at different excitation densities with the light sums stored at the corresponding excitation current densities. While the value of the light sums increases slightly (by a factor of 1.5 to 3) and in the limiting case of long-duration excitation remains constant, the luminescent brightness increases rapidly with increasing current density (by about two orders of magnitude in the density range from 10^{-8} to 10^{-6} A/cm^2; see, for instance, Fig. 28).

Consequently, after the cessation of the excitation the number of electrons which recombine from very shallow trapping levels or directly from the conduction band increases with j during the decay process. The electrons localized in deeper trapping levels (whose number remains almost constant) at higher values of j have a smaller effect on the luminescence intensity directly after the cessation of the excitation, because during this stage the luminescence is emitted from the shallow levels, whose degree of occupation increases with increasing excitation density.

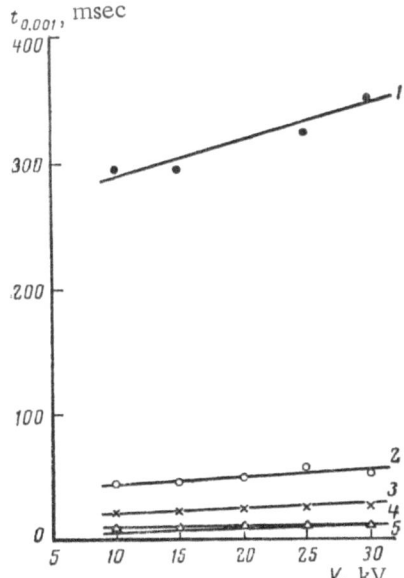

Fig. 59. Dependence of the decay time to 0.1% of the original brightness on the initial electron energy. 1) ZnS-Ag luminophor; 2,3) different specimens of the ZnS·CdS-Ag luminophor; 4) ZnS-Tu luminophor; 5) $Sr_3(PO_4)_2$-Eu.

The relation between the reduced importance of the electrons localized in the principal level and the reduction in the decay time to 0.001 of the original brightness is clearly evident in Fig. 56. The dashed curves in the figure show the ratio S/B, where S is the light sum in the principal level and B is the luminescent brightness at the instant of excitation. This ratio characterizes qualitatively the participation in the radiation process of the electrons localized in the principal level and of the electrons released from very shallow levels; the larger the ratio, the smaller the role of the electrons from the deeper levels. As is evident from Fig. 56, this ratio changes greatly as the current density is increased.

The solid curves give τ, which is the relative duration of the decay to 0.001 of the original brightness. Both these curves have a similar shape along an appreciable part of their length, i.e., the reduction in the number of electrons originating from deep levels results in a decrease of the afterglow time.

To derive a more accurate quantitative relation between τ and S/B, we investigated in detail a specimen of the ZnS·CdS-Ag luminophor with one peak on the t.l.c. at about $-146°$C.

In addition to determining the light sum from the t.l. curves for this specimen, we calculated the light sums from the decay, as the area under a decay curve plotted in the normal coordinates. Since the decay measurements were made over a wide range of variation of the intensity and time (approximately four orders of magnitude), to calculate the light sums by this method the decay curves were constructed on two scales (Fig. 57a and b).

The nature of the variation of the light sums calculated from the t.l.c. ($S_{t.l.}$) and from the decay (S_{decay}) is the same. Comparing the variation of the decay time to 0.001 of the original brightness (τ) with the variation of the quantities $S_{t.l.}/B$ and S_{decay}/B, again we obtain for these quantities curves of a similar shape (Fig. 58). A slightly larger difference is observed in the case of the ratio $S_{t.l.}/B$.

The described difference between the steady state brightness, the occupation of levels and the decay time may influence the results obtained during measurements of the decay time as a function of the excitation conditions: 1) excitation density, 2) duration of excitation, 3) excitation mode (pulsed or continuous).

The small value of the light sum stored in trapping levels (estimates from the peak intensity during thermoluminescence show that it hardly reaches 0.05% of the original brightness upon excitation) and especially its rapid saturation confirm that the process of ejection of electrons from the levels by fast electrons and thermal vibrations of the lattice is intense. These two actions may exert a much greater influence as the cathode beam intensity increases.

As the accelerating voltage increases, the duration of the afterglow is observed to increase slightly (Fig. 59). This is related to the reduction of the excitation density per unit volume, resulting from a deeper penetration of the electron beam into the luminophor and a more uniform distribution of energy over the penetration of the electron beam.

§6. Relation between the Luminescence Time and the Position and Occupation of Trapping Levels

As has been noted, the inertial properties of luminophors are determined by the distribution of localized trapping levels. To establish the relation between the luminescence time and decay mode and the position and occupation of the trapping levels in a broad class of luminophors, we measured the t.l.c. and compared them

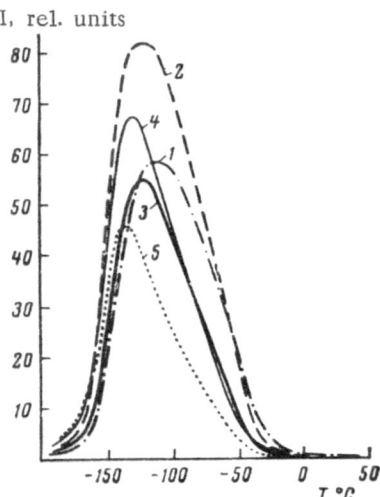

I, rel. units

Fig. 60. The t.l. curves of the ZnS·CdS-Ag luminophors with different contents of CdS (in %). 1) 0; 2) 2.5; 3) 5; 4) 7.5; 5) 10.

with the decay curves of the same luminophors. A close relation between the intrinsic decay during the later stages and the distribution of electron traps has been observed in [74, 75].

Using our method for measuring the decay time, we were able to broaden appreciably the time interval investigated in the direction of shorter times (to 10^{-5} sec). Consequently, it was found possible to examine the effect of very shallow trapping levels on the decay mode during the initial stages.

In the luminescence theory of crystal phosphors it is usually assumed that the probability p of releasing electrons from traps is given by the formula

$$p = p_0 e^{-\frac{E}{kT}},$$

where E is the energy depth of a trap below the conduction band.

Thus the probability p depends strongly on the depth E of the trapping level. If we assume that $p_0 = 10^9$ sec^{-1}, then at 18°C for traps characteristic of the luminophors investigated the release time is

E, eV ~	0.25	0.27	0.40	0.52
$\theta = \frac{1}{p}$, sec ~	$2 \cdot 10^{-5}$	$4 \cdot 10^{-5}$	$6 \cdot 10^{-3}$	$7 \cdot 10^{-1}$

It is evident from these figures how rapidly the electron release probability decreases with increasing depth of traps.

The predicted increase of the decay time with increasing energy depth of trapping levels is well confirmed experimentally in the case of the series of specimens of the ZnS·CdS-Ag luminophor with different contents of CdS.

As is known, the introduction of CdS into zinc sulfide results in a shift of the radiation and absorption bands in the direction of longer wavelengths. This is due to the replacement of one ion in the host material with another ion from the same periodic group, but exhibiting a different polarization of the electron shell, accompanied by a narrowing of the forbidden band. In this case the trapping levels should move closer to the conduction band, i.e., they become shallower. This in turn results in a higher probability of electron release from traps and a reduction of the afterglow time.

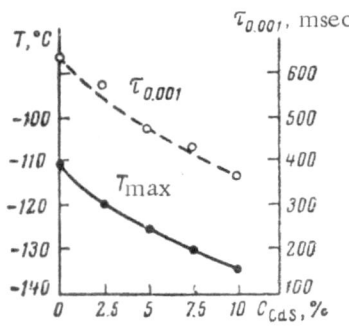

Fig. 61. Shift of the t.l.c. maximum (solid curve) and the variation of the decay time to 0.001 of the original brightness (dashed curve) of specimens of the ZnS·CdS-Ag luminophor with different contents of CdS.

Actually, an increase in the percentage content of CdS from 0 to 10% in specimens of the luminophor mentioned results in a smooth shift of the peaks on the thermoluminescence curves into the region of lower temperatures (Fig. 60). The afterglow time of the specimens also decreases correspondingly (Fig. 61).

The fact that the decay time increases as the energy depth of the level increases has been confirmed experimentally for specimens of different classes. As is known, depending on the preparation conditions, in the ZnS phosphors we can have levels of different depths [71], corresponding to the introduction of oxygen and activators and to the formation of other lattice imperfections. Figure 62 shows the t.l.c. of ZnS-Tu, ZnS·CdS-Ag and ZnS-Ag. In accordance with the

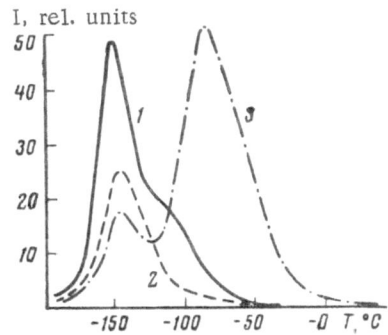

I, rel. units

Fig. 62. The t.l. curves of lumino-phors. 1) ZnS-Tu; 2) ZnS·CdS-Ag; 3) ZnS-Ag.

position of the t.l. peaks the decay time to 0.001 of the original value in ZnS-Tu was 10 msec (curve 1), in ZnS·CdS-Ag 40 msec (curve 2), and in ZnS-Ag 450 msec (curve 3). Also, since the fundamental peak of the ZnS·CdS-Ag phosphor is lower than that of ZnS-Tu, curve 2 intersects curve 1 at +50°C. Consequently, during the later stages the decay of this phosphor is slower and the reduction of the luminescence intensity to 0.001 of the original value takes a longer time than in the ZnS-Tu phosphor.

In a number of cases the position of the t.l. peaks for a series of phosphors differing in composition or preparation technique may remain the same, but the dispersion and the relative development of levels of different depth change appreciably. This situation exists, in particular, in the ZnS-Ag, Al phosphors with different contents of Al.

In such series of phosphors the development of shallow levels, corresponding to the low-temperature part of the t.l.c., increases the brightness during the intitial stages of decay, i.e., it shortens the decay time to 0.001 (or some other small value); conversely, an increase in the number of deep levels slows down the afterglow process and increases the brightness during the later stages of decay, with the resulting increase in the decay time to 0.001.

We investigated a group of specimens of the ZnS-Ag luminophor with different contents of Al acting as a coactivator. The Al concentration was varied between 7.2×10^{-6} and 2.4×10^{-4} g/g. The thermoluminescence curves of these specimens are shown in Fig. 63. All the specimens investigated exhibited one peak on the thermoluminescence curve with a maximum close to $-106°$C.

The relative changes of brightness at the instant of excitation and of the stored light sums determined by the t.l. method are shown in Fig. 64a. It should be noted that this series of luminophors exhibits a much longer decay time to 0.001 of the original brightness than the specimens of the ZnS·CdS-Ag and ZnS-Tu luminophors discussed in §5. Their lifetimes ($\tau_{0.001}$) for increasing concentrations of Al are: 100 msec for 7.2×10^{-6} g/g, 390 msec for 2.2×10^{-5} g/g, 620 msec for 7.2×10^{-5} g/g, and 590 msec for 2×10^{-4} g/g.

Consequently, a satisfactory correspondence between the variation of $\tau_{0.001}$ and $S_{t.l.}/B$ (Fig. 64b) is obtained only for the values of $S_{t.l.}$ pertaining to the high-temperature part of the t.l.c. In Fig. 64a the light sum was determined as the area under the descending high-temperature part of the t.l.c. contained in the interval from -60 to $+40°$C. This again confirms the fact that in all slowly decaying phosphors the system of deep levels plays the most important role during the later stages of decay.

The described relation between the duration of the luminescence and the number of trapping levels in the main system which takes part in the decay of the phosphor is completely understood. As the number of these levels decreases, an increasing number of electrons recombine either directly from the conduction band or after they have been captured in shallower levels. The recombination of these electrons raises the luminescence level at the instant of excitation and thus increases the rate of brightness decay to 0.001 of the original value. This is achieved at a cost of lower depletion of the trapping levels.

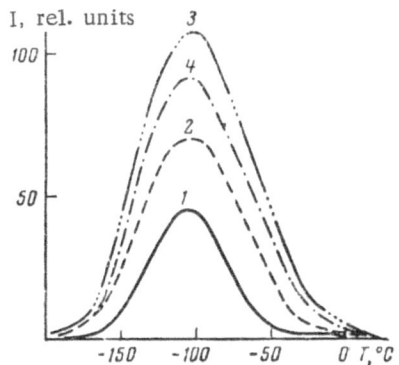

I, rel. units

Fig. 63. The t.l. curves of the ZnS-Ag, Al luminophors with different contents of Al. 1) 7.2×10^{-6} g/g; 2) 2.2×10^{-5} g/g; 3) 7.2×10^{-5} g/g; 4) 2×10^{-4} g/g.

§7. Dependence of the Duration of Luminescence on Temperature

To investigate the nature of the variation of the inertial properties of the luminophor specimens on the temperature, we measured the decay time of these specimens in the temperature range from

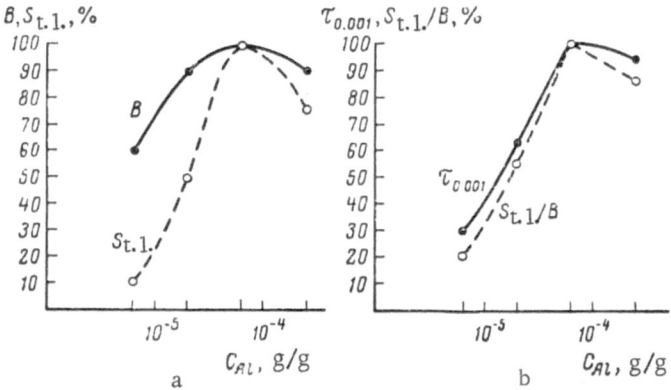

Fig. 64. Relative change of the brightness B at the instant of excitation and of the light sum $S_{t.1.}$ of the high-temperature part of the t.l.c. (a), and of the decay time $\tau_{0.001}$ of the original brightness and of the ratio $S_{t.1.}/B$ (b) in ZnS-Ag, Al.

−70 to +60°C. The apparatus used for these measurements was described in detail in §4.

As has been shown in a number of papers (see, for instance, [79]), depending on the degree of vacuum, the temperature of the phosphor may differ greatly from the temperature of the heat-conducting substrate. We took this into account during our temperature measurements.

Before the measurements were made, the phosphor was maintained for a sufficiently long time at the required temperature under low vacuum so that its temperature would become almost the same as that of the metallic substrate. The decay was measured at a voltage of 15 kV and a current density of 1×10^{-7} A/cm².

Figure 65 shows the curves of the decay time of the afterglow as a function of temperature between −70 and +60°C for the ZnS-Ag and ZnS-Tu luminophors. Separate curves are drawn for the times required to decay to 0.1, 0.01, and 0.001 of the original brightness observed at the instant of excitation with the electron beam. The first series of curves establishes the effect of temperature on the initial stages of decay, and the following series describe the influence of the temperature on the duration of a more complete decay to 0.01 and 0.001 of the original brightness.

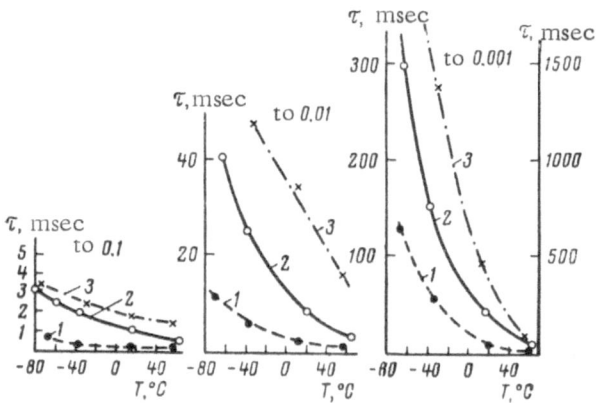

Fig. 65. Dependence of the decay time to 0.1, 0.01, and 0.001 of the original brightness on the temperature. 1) ZnS-Tu; 2,3) various specimens of ZnS-Ag. The right scale is for curve 3 (to 0.001).

118

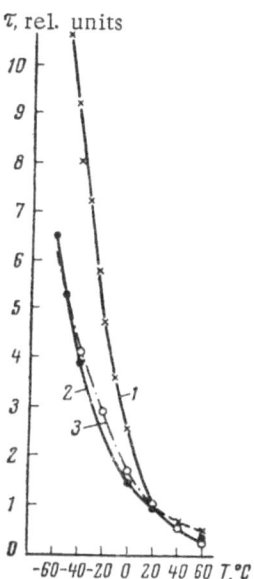

Fig. 66. Dependence of the relative decay time to 0.001 of the original brightness on the temperature (the value of $\tau_{0.001}$ at 20°C is taken as unity). 1) ZnS-Tu; 2,3) some specimens of ZnS-Ag.

The early stages of the decay process up to 0.03 msec are subject to small temperature variations, because they are related to the depletion of very shallow levels for which even −70°C is a high temperature. Analysis of these curves shows that for all luminophors investigated the decay time to 0.1 and 0.01 of the initial brightness is reduced greatly as the temperature is increased; upon cooling it increases.

The following characteristic features of the process should be noted.

The relative lengthening of the decay time with decreasing temperature becomes more pronounced as we go successively to later stages of decay. The maximum change of the decay time occurs for the decay to 0.001 of the original brightness. This is clearly evident from Fig. 66, in which the decay times up to that limit are reduced to the temperature of +20°C.

The luminophor activated with Tu exhibits a large variation of decay time; the duration of its afterglow to 0.001 of the original brightness even at 0°C increases by a factor of 2.5 relative to the duration of its afterglow at room temperature.

The nature of the temperature variation of the decay time is explained by the distribution of localized levels of the given specimens of luminophors (see Fig. 62, 1). This peculiarity of the ZnS-Tu luminophor is related directly to the presence of a second very deep trapping level, whose reduced rate of depletion at lower temperatures results in a sharp lengthening of the afterglow.

SUMMARY

In the present paper we analyzed the main properties of the cathodoluminescence, which reflect the specific characteristics of this type of luminescence. We investigated the luminescence efficiency and its dependence on the excitation conditions, and the luminescence inertia and its relation to the development and location of electron localization sites. The following results were obtained.

1. The various possible types of excitation energy losses were analyzed.

2. A detailed theoretical analysis of the energy losses due to the thermal stabilization of electrons and holes was made and it was shown that during the cathodoluminescence unavoidable losses of this type are very large; they reduce the luminescence efficiency by a factor of 2.5.

3. From the estimates of other types of losses we concluded that the maximum luminescence efficiency during the cathodoluminescence cannot exceed 0.27 to 0.30.

4. The transmission of the electron beam through sublimated layers of zinc sulfide luminophors was investigated in detail. This allowed us to establish the dependence of the electron penetration depth on the voltage, and also the energy losses at different electron penetration depths in the luminophor layer.

5. The dependence of the spectral composition, brightness, and luminescence energy yield on the current density, voltage, and temperature was investigated in a number of zinc sulfide and phosphate luminophors. The luminescence efficiency was calculated to be 0.256 for one luminophor of the ZnS-Ag type.

6. The luminescence decay of various types of cathodoluminophors to 0.1, 0.01, and 0.001 of the original brightness was investigated. It was found that there are two luminescence processes that differ greatly in duration and are superimposed on each other. The reasons for the shorter afterglow at higher excitation densities were analyzed. The position and development of trapping levels in the luminophors were investigated by the thermoluminescence method, and it was established that there is a direct relation between the decay mode and the depletion of levels at a certain depth. The effect of temperature on these processes was considered.

LITERATURE CITED

1. L. N. Dobretsov and V. A. Matskevich, Zhur. Tekh. Fiz. 32:734, 1957.
2. V. A Matskevich, Zhur. Tekh. Fiz. 32:289, 1957.
3. L. D. Landau and E. M. Livshits, Electrodynamics of Continuous Media, Gostekhizdat, 1957.
4. O. N. Krokhin and Yu. M. Popov, Zhur. Eksptl. i Teoret. Fiz. 38:1589, 1960.
5. B. Davydov and I. Shmushkevich, Zhur. Eksptl. i Teoret. Fiz. 10:1043, 1940.
6. N. R. Whetten and A. B. Laponsky, Phys. Rev. 107:152, 1957.
7. M. T. Gornyi, Zhur. Eksptl. i Teoret. Fiz. 35:281, 1958.
8. V. L. Levshin and V. F. Tunitskaya, Optika i Spektroskopiya 9:223, 1960.
9. M. N. Alentsev and E. I. Panasyuk, Optika i Spektroskopiya 5:207, 1958.
10. L. Drozd and V. L. Levshin, Optika i Spektroskopiya 11:648, 1961.
11. L. R. Koller and E. D. Alden, Phys. Rev. 83:684, 1951.
12. W. Ehrenberg and J. Franks, Proc. Phys. Soc. 66:1057, 1953.
13. J. R. Yong, J. Appl. Phys. 27:1, 1956.
14. J. R. Yong, J. Appl. Phys. 28:524, 1957.
15. C. Feldman, Phys. Rev. 117:455, 1960.
16. A. Ya. Vyatskin and A. F Makhov, Zhur. Eksptl. i Teoret. Fiz. 28:740, 1958.
17. A. Ya. Vyatskin and A. F. Makhov, Fizika Tverdogo Tela 2:885, 1960.
18. F. J. Studer, D. A. Cusano, and A. U. Joung, J. Opt. Soc. Am. 41:559, 1951; F. J. Studer and D. A. Cusano, J. Opt. Soc. Am. 45:493, 1955.
19. F. D. Klement and Ya. M. Zelikin, Author's certificate No. 11201, 1950.
20. E. I. Blazhnova and G. A. Kruglova, Reports of the Seventh Conference on Luminescence (Crystallophosphors), Moscow, 1958, Tartu, p. 373, 1959.
21. K. V. Shalimova, Doklady Akad. Nauk SSSR 80:587, 1951.
22. C. Feldman and M. O'Hara, J. Opt. Soc. Am. 47:300, 1957; J Opt. Soc. Am. 48:816, 1958.
23. N. A. Vlasenko, Reports of the Seventh Conference on Luminescence (Crystallophosphors), Moscow, 1958, Tartu, p. 365, 1959; Optika i Spektroskopiya 8:414, 1960.
24. L. R. Koller and H. D. Coghill, J. Electrochem. Soc. 107:973, 1960.
25. V. V. Golubets, Reports of the Seventh Conference on Luminescence (Crystallophosphors), Moscow, 1958, Tartu, p. 353, 1959.
26. É. Ya. Arapova, Reports of the Seventh Conference on Luminescence (Crystallophosphors), Moscow, 1958, Tartu, p. 358, 1959.
27. É. Ya. Arapova, Izvest. Akad. Nauk SSSR, Ser. Fiz. 25:324, 1961.
28. H. Sacamoto and S. Tanaka, J. Appl. Phys. Japan 29:412, 1960.
29. P. Goldberg, D. Bracco, and A. Kremheller, J. Electrochem. Soc. 102:79C, 1955.
30. É. Ya. Arapova, Reports of the Ninth Conference on Luminescence (Crystallophosphors), Kiev, p. 323, 1960.
31. A. G. Zavrazhin and A. I. Blazhevich, Reports of the Seventh Conference on Luminescence (Crystallophosphors), Moscow, 1958, Tartu, p. 316, 1959.
32. B. I. Pochtarev, K. K. Raspletin, and D. V. Fetisov, Izvest. Akad. Nauk SSSR, Ser. Fiz. 32:462, 1959.
33. B. I. Pochtarev, Izvest. Akad. Nauk SSSR, Ser. Fiz. 25:514, 1961.
34. Yu. V. Voronov and A. G. Ovchinnikov, Pribory i Techn. Eksperim. (in press).
35. Physics and Procedures of Spectral Analysis, Fizmatgiz, p. 117, 1961.

36. S. A. Fridman and V. V. Shchaenko, Reports of the Seventh Conference on Luminescence (Crystallophosphors), Moscow, 1958, p. 288, 1959.

37. P. P. Feofilov, Reports of the Fifth Conference on Luminescence (Crystallophosphors), Tartu, p. 3, 1957.

38. A. A. Manenkov, A. M. Prokhorov, Z. A. Trapeznikova, and M. V. Fok, Izvest. Akad. Nauk SSSR, Ser. Fiz. 21:779, 1957.

39. I. V. Stepanov and P. P. Feofilov, Doklady Akad. Nauk SSSR 108:615, 1956.

40. Z. A. Trapeznikova, Zhur. Eksptl. i Teoret. Fiz. 21:283, 1951.

41. V. L. Levshin, É. Ya. Arapova, and E. G. Baranova, Trudy Komissii Anal. Khim. 12:393, 1960.

42. V. P. Nazarova, Izvest. Akad. Nauk SSSR, Ser. Fiz. 25:332, 1961.

43. A. V. Moskvin, Cathodoluminescence, GITTL, 1948.

44. S. Rotshild, Materials of the Inst. Conf. held in Brussels 4:705, 1958.

45. M. V. Fok and L. A. Vinokurov, Optika i Spektroskopiya 4:118, 1958.

46. Z. A. Trapeznikova and V. V. Shchaenko, Doklady Akad. Nauk SSSR 106:230, 1956.

47. N. Riehl and M. Schön, Z. Physik 114:682, 1939.

48. F. Möglich and R. Rompe, Physik. Z. 41:552, 1940.

49. R. P. Johnson, Phys. Rev. 55:881, 1939.

50. V. L. Levshin and B. D. Ryzhikov, Optika i Spektroskopiya 12:3, 1962.

51. V. L. Levshin, Yu. V. Voronov, V. B. Gutan, S. A. Fridman, and V. V. Shchaenko, Izvest. Akad. Nauk SSSR, Ser. Fiz. 25:392, 1961.

52. V. L. Levshin, Yu. V. Voronov, V. B. Gutan, S. A. Fridman, and V. V. Shchaenko, Optika i Spektroskopiya, Collection of Papers 1:230, 1963.

53. G. R. Fonda, J. Opt. Soc. Am. 43:561, 1939.

54. A. Bril, H. Klasens, and T. J. Westerhof, Physica 24:821, 1958.

55. P. Lenard and S. Saeland, Ann. Phys. 28:476, 1909.

56. W. B. Nottingham, J. Appl. Phys. 10:73, 1939.

57. W. B. Nottingham, J. Appl. Phys. 8:877, 1937.

58. J. W. Strange and S. T. Henderson, Proc. Phys. Soc. 58:369, 1946.

59. A. Bril, Physica 15:361, 1949.

60. A. Bril and F. A. Kröger, Philips Tech. Rev. 12:120, 1960.

61. K. Yanchevskii, Izvest. Akad. Nauk SSSR 9:463, 1945.

62. G. J. Garlick, Brit. J. Appl. Phys. Supplement 20:5103, 1955.

63. M. C. Kay, Advances in Electronics 1:65, 1949.

64. C. G. Hill, Brit. J. Appl. Phys. Supplement 4:6, 1955.

65. G. J. Garlick, Brit. J. Appl. Phys. 3:162, 1952.

66. A. Bril and H. A. Klasens, Philips Tech. Rev. 7:401, 1952.

67. A. Bril and H. A. Klasens, Philips Tech. Rev. 15:63, 1953.

68. V. L. Levshin, Optika i Spektroskopiya 11:362, 1961.

69. M. L. Kats and R. E. Solomonyuk, Doklady Akad. Nauk SSSR 24:682, 1934; N. A. Brilliantov and Z. L. Morgenshtern, Zhur. Eksptl. i Teoret. Fiz. 8:401, 1938.

70. Ch. B. Lushchik, Trudy Inst. Fiz. i Astron., Akad. Nauk Est.SSR (3):1955.

71. V. L. Levshin, V. F. Tunitskaya, and A. A. Cherepnev, Optika i Spektroskopiya 1:255, 1956; V. L. Levshin and V. F. Tunitskaya, Optika i Spektroskopiya 2:354, 1957.

72. J. T. Randall and M. H. F. Wilkins, Proc. Roy. Soc. A 184:347, 1945.

73. I. A. Parfianovich, Zhur. Eksptl. i Teoret. Fiz. 26:696, 1954.

74. N. V. Zhukova, G. K. Evdokimova, and L. V. Levshin, Izvest. Akad. Nauk SSSR, Ser. Fiz. 25:476, 1961.

75. V. L. Levshin, Izvest. Akad. Nauk SSSR, Ser. Fiz. 26:450, 1962.

76. A. G. Zavrazhin, A. I. Blazhevich, and A. V. Lavrov, Optika i Spektroskopiya 8:550, 1960.

77. Yu. M. Popov, Optika i Spektroskopiya 7:697, 1959.

78. V. L. Levshin, V. B. Gutan, and É. N. Karzhavina, Optika i Spektroskopiya 6:372, 1959.

79. L. L. Katan, Vacuum 1:191, 1951.

BIBLIOGRAPHY

Bibliography of Work Performed in the Luminescence Laboratory of the P. N. Lebedev Physics Institute of the USSR Academy of Sciences* between 1934 and 1961†

Many of the journals cited in this bibliography are available in cover-to-cover English translation. The names, publishers, and year in which the translation began are listed below.

Russian Title	English Title	Publisher	Year
Doklady Akademii Nauk SSSR	Proceedings of the Academy of Sciences of the USSR, Section: Chemistry	Consultants Bureau	1956
	Proceedings of the Academy of Sciences of the USSR, Section: Physical Chemistry	Consultants Bureau	1957
	Soviet Physics — Doklady	American Institute of Physics	1956
Zhurnal tekhnicheskoi fiziki	Soviet Physics — Technical Physics	American Institute of Physics	1956
Zhurnal éksperimental'noi i teoreticheskoi fiziki	Soviet Physics — JETP	American Institute of Physics	1955
Zhurnal fizicheskoi khimii	Russian Journal of Physical Chemistry	Chemical Society (London)	1959
Uspekhi fizicheskikh nauk	Soviet Physics — Uspekhi (partial translation)	American Institute of Physics	1958
Izvestiya Akademii Nauk SSSR: Seriya fizicheskaya	Bulletin of the Academy of Sciences of the USSR: Physical Series	Columbia Technical Translations	1954
Zavodskaya laboratoriya	Industrial Laboratory	Instrument Society of America	1958
Izvestia Akademii Nauk SSSR: Otdelenie khimicheskikh nauk	Bulletin of the Academy of Sciences of the USSR: Division of Chemical Science	Consultants Bureau	1952

---

* Compiler — senior bibliographer of the FIAN library N. V. Slesareva. Editors: Doctor of Physical and Mathematical Sciences Professor V. L. Levshin and Candidate of Technical Sciences S. A. Fridman.
† The papers marked with an asterisk are from workers of the Luminescence Laboratory working in other organizations.

Russian Title	English Title	Publisher	Year
Fizika tverdogo tela	Soviet Physics — Solid State	American Institute of Physics	1959
Zhurnal prikladnoi khimii	Journal of Applied Chemistry USSR	Consultants Bureau	1950
Optika i spektroskopiya	Optics and Spectroscopy	American Institute of Physics	1959
Zhurnal analiticheskoi khimii	Journal of Analytical Chemistry	Consultants Bureau	1952
Zhurnal neorganicheskoi khimii	Journal of Inorganic Chemistry	Chemical Society (London)	1959

1934

V. V. Antonov-Romanovskii, Natural damping of zinc phosphors in single crystals, * Dokl. Akad. Nauk SSSR 3(6):432, 1934.

V. V. Antonov-Romanovskii and V. L. Pul'ver, Calculation of a searchlight beam from the aberation characteristics of the reflector, * Zhur. Tekh. Fiz. 4(3):568, 1934.

E. M. Brumberg and S. I. Vavilov, Statistical structure of an interference field, * Doklady Akad. Nauk SSSR 3(5):322, 1934; German translation, Doklady Akad. Nauk SSSR 3(5):325, 1934.

E. M. Brumberg and S. I. Vavilov, The accuracy in the photometric method of quenching, * Doklady Akad. Nauk SSSR 3(6):405, 1934; German translation, Doklady Akad. Nauk SSSR 3(6):408, 1934.

V. L. Levshin and V. V. Antonov-Romanovskii, Investigations into phosphorescence. I. The hyperbolic law of damping of phosphors, * Zhur. Eksptl. i Teoret. Fiz. 4(10):1022, 1934; German translation, Physik. Z. Sowjetunion 5(6):796, 1934.

V. L. Levshin, V. V. Antonov-Romanovskii, and L. A. Tumerman, Investigations into phosphorescence. II. A study of the phenomenon of quenching of phosphorescence by infrared rays in order to apply it to photography in the infrared part of the spectrum, * Zhur. Eksptl. i. Teoret. Fiz. 4(10):1033, 1934; German translation, Physik. Z. Sowjetunion 5(6):811, 1934.

V. L. Levshin, The connection between absorption and luminescence spectra in concentrated solutions of dyes, * Doklady Akad. Nauk SSSR 2(7):405, 1934 (preliminary communication); French translation, Compt. rend. acad. sci. URSS 2(7):408, 1934.

A. A. Schischlovski and S. I. Vavilov, Die Abklingungsgesetze des Phosphoreszens von Farbstofflösungen,* Physik. Z. Sowjetunion 5(3):379, 1934.

S. I. Vavilov, State Optics Institute, * in: Scientific and Technical Maintenance in Heavy Industry, Moscow-Leningrad, ONTI, pp. 30-40, 1934. (Review)

S. I. Vavilov, Possible reasons for the blue γ-luminescence of liquids, Doklady Akad. Nauk SSSR 2(8):457, 1934; German translation, Doklady Akad. Nauk SSSR 2(8):459, 1934.

S. I. Vavilov, Light (Review), Technical Encyclopedia, Vol. 20, p. 281, 1935.

S. I. Vavilov, Über die Abklingungsgesetze der umkehrbaren Lumineszenzzerscheinungen, Physik. Z. Sowjetunion 5(3):369, 1934.

S. I. Vavilov and E. Brumberg, Visuelle Messungen der statistischen Photonenschwankungen, * Trudy GOI, 10(95):919, 1934.

1935

V. V. Antonov-Romanovskii, The effect of the uneven distribution of phosphorescence centers and related factors on the damping of Lenard phosphors, Doklady Akad. Nauk SSSR (2):105, 1935.

V. V. Antonov-Romanovskii, Damping of zinc phosphors in single crystals, * Zhur. Fiz. Khim. 6(8):1022, 1935; German translation, Physik. Z. Sowjetunion 7(3):366, 1935.

M. A. Konstantinova-Shlezinger, Photochemical decomposition of sulfur dioxide, Zhur. Fiz. Khim. 6(5):601, 1935.

M. A. Konstantinova-Schlezinger, Eine neue Fluoreszenmethode zur Bestimmung von geringen ozonkonzentrationen, Acta Physicochim. URSS 3(4):435, 1935.

V. L. Levshin and S. A. Fridman, An investigation of the luminescence of ZnS-Cu-α ZnS-CdS-Cu-α phosphors and the structure of phosphorescence centers, * Zhur. Fiz. Khim. 6(10):1277, 1935.

V. L. Levshin and M. N. Alentsev, An investigation of the phosphorescence of calcites, * Doklady Akad. Nauk SSSR 2(1):54, 1935; French translation, Compt. rend. acad. sci. URSS 2(1):56, 1935.

V. L. Levshin, The connection between absorption and luminescence spectra in concentrated solutions of dyes, * Zhur. Fiz. Khim. 6(1):1, 1935; English translation, Acta Physicochim. URSS 1(5):685, 1935.

V. L. Levshin, The connection between absorption and luminescence spectra in weak solutions of dyes, Doklady Akad. Nauk SSSR 1(7-8):474, 1935 (preliminary communication); German translation, Compt. rend. acad. sci. URSS 1(7-8):479, 1935.

V. L. Levshin, Correspondence between absorption and luminescence spectra in weak solutions of dyes (effect of temperatures and solvent),* Zhur. Fiz. Khim. 6(8):991, 1935; English translation, Acta Physicochim. URSS 2(2):221, 1935.

S. I. Vavilov, Physics,Pod. znam. marks. (1):124, 1935.

S. I. Vavilov, Physical Optics of Leonard Euler (1707-1783), collection of articles in honor of the 150th anniversary of his death, Moscow-Leningrad, Izd. AN SSSR, p. 29, 1935.

S. I. Vavilov, A photometric method of quenching and its use, Priroda (12):8, 1935.

1936

V. V. Antonov-Romanovskii, Direct proof of the bimolecular scheme of luminescence of zinc phosphors, Doklady Akad. Nauk SSSR 2(3):93, 1936; French translation, Compt. rend. acad. sci. URSS 2(3):97, 1936.

V. V. Antonov-Romanovskii, N. Grigorov, N. Dobrotin, and I. Frank, Work with the Wilson chamber in 1935, in: Transactions of the Elbrus Expedition of the Academy of Sciences of the USSR and VIÉM, 1934 and 1935, Moscow-Leningrad, Izd AN SSSR, p. 37, 1936.

M. A. Konstantinova-Shlezinger, A new fluorescence method for the determination of small concentrations of ozone, in: Trudy FIAN,Vol. 1, No. 1, Moscow-Leningrad, Izd. AN SSSR, p. 119, 1936.

M. A. Konstantinova-Shlezinger, A fluorescence method for the determination of ozone in air, in: Transactions of the Elbrus Expedition of the Academy of Sciences of the USSR and VIÉM, 1934 and 1935, Moscow-Leningrad, Izd. AN SSSR, p. 49, 1936.

V. L. Levshin, New trends in Soviet physics (March session of the Academy of Sciences of the USSR), Vestnik Akad. Nauk SSSR (4-5):62, 1936.

V. L. Levshin and M. N. Alentsev, Experience in the introduction of quantitative methods of investigation in the luminescence analysis of minerals, * Trudy MGRI 1:207, 1936.

V. L. Levshin, Luminescing Compositions, Moscow-Leningrad, Izd. AN SSSR, 134 pp., 1936.

V. L. Levshin, Phosphorescence, Bol'shaya Sov'etskaya Éntsiklopediya, First Edition, Vol. 58, Moscow, p. 273, 1936; Physics Dictionary, Moscow, ONTI, Vol. 5, p. 470, 1939.

V. L. Levshin, Recherches sur la décroissance de la luminescence et le mécanisme d'émission de différentes substances, Acta Phys. Polon. 5:301, 1936.

S. I. Vavilov, Yield and duration of fluorescence, * Doklady Akad. Nauk SSSR 3(6):271, 1936; German translation, Compt. rend. acad. sci. URSS 3(6):271, 1936.

S. I. Vavilov, Optical work and views of M. V. Lomonosov, * Priroda (12):121, 1936.

S. I. Vavilov, Trends in the development of the Optical Institute (Report), * Izvest. Akad. Nauk SSSR, Ser. Fiz. (1-2):163, 1936; Uspekhi Fiz. Nauk 16(7):872, 1936.

S. I. Vavilov and A. N. Sevchenko, Quenching of fluorescence by the solvent, * Doklady Akad. Nauk SSSR 3(6):277, 1936; German Translation, Compt. rend. acad. sci. URSS 3(6):277, 1936.

S. I. Vavilov, Die Auslöschung der Fluoreszenz in flüssigen Lösungen, Acta Phys. Polon. 5:417, 1936.

S. I. Vavilov, Light fluctuations and their measurement by a visual method, in: Transactions of the First Conference on Physiological Optics, December 25-29, 1934, Moscow-Leningrad, Izd. AN SSSR, p. 332.

L. A. Vinokurov and V. L. Levshin, An investigation of the quenching of phosphors activated with organic substances, Doklady Akad. Nauk SSSR 2(4):133, 1936 (preliminary communication); French translation, Compt. rend. acad. sci. URSS 2(4):135, 1936.

L. A. Vinokurov and V. L. Levshin, An investigation into the quenching of boron and aluminum phosphors, Zhur. Fiz. Khim. 8(2):181, 1936; English translation, Physik. Z. Sowjetunion 10(1):10, 1936.

1937

V. V. Antonov-Romanovskii, The law of Quenching of Phosphors (thesis for the degree of Candidate of Physical and Mathematical Sciences), in: Trudy FIAN, Vol. 1, No. 2, Moscow-Leningrad, Izd. AN SSSR, p. 35, 1937.

V. V. Antonov-Romanovskii, Quantitative measurements of the quenching of luminescence of zinc phosphor at various temperatures, Doklady Akad. Nauk SSSR 17(3):95, 1937; French translation, Compt. rend. acad. sci. URSS 17(3):95, 1937.

M. A. Konstantinova-Shlezinger, Determination of ozone in air at a height of 9620 m by a fluorescence method, Doklady Akad. Nauk SSSR 14(4):187, 1937; French translation, Compt. rend. acad. sci. URSS 14(4):187, 1937.

M. A. Konstantinova-Shlezinger, Results of the determination of ozone in air by a fluorescence method, Izvest. Akad. Nauk SSSR, Ser. Fiz. (2):213, 1937.

V. L. Levshin, An investigation into absorption and luminescence spectra of uranyl salts and their solutions, Izvest. Akad. Nauk SSSR, Ser. Fiz. (2):185, 1937; English translation, Acta Physicochim. URSS 6(5):661, 1937.

V. L. Levshin and S. N. Rzhevkin, The mechanism of the luminescence of liquids under the action of ultrasonics, Doklady Akad. Nauk SSSR 16(8):407, 1937; English translation, Compt. rend. acad. sci. URSS 16(8):399, 1937.

V. L. Levshin, Luminescence analysis, Physics Dictionary, Vol. 3, Moscow, ONTI, p. 375, 1937.

V. L. Levshin, An attempt at a quantum interpretation of the mirror symmetry of absorption and luminescence spectra, Zhur. Fiz. Khim. 9(1):1, 1937; English translation, Acta Physicochim. URSS 6(2):213, 1937.

S. I. Vavilov, P. G. Glukhov, and I. A. Khvostikov, The depolarization of fluorescence of solutions at high concentrations, Doklady Akad. Nauk SSSR 16(5):267, 1937; German translation, Compt. rend. acad. sci. URSS 16(5):251, 1937.

S. I. Vavilov, Comments on the molecular viscosity of liquids, Izvest. Akad. Nauk SSSR, Ser. Fiz. (3):345, 1937; English translation, Acta Physicochim. URSS 7(1):49, 1937.

S. I. Vavilov, A method for the determination of the true polarization of fluorescence of solutions at high concentrations, Doklady Akad. Nauk, SSSR 16(5):263, 1937; German translation, Compt. rend. acad. sci. URSS 16(5):255, 1937.

S. I. Vavilov, Optical work and views of M. V. Lomonosov, Izvest. Akad. Nauk SSSR, OON (1):235, 1937.

S. I. Vavilov, The nature of elementary radiators and interference phenomena, Doklady Akad. Nauk SSSR 17(9):459; English translation, Compt. rend. acad. sci. URSS 17(9):463, 1937.

S. I. Vavilov, Physics in the creative genius of D. I. Mendeleev, in: Transactions of the Mendeleev Jubilee Conference, Vol. 2, Moscow-Leningrad, Izd. AN SSSR, p. 8, 1937; German translation, p. 13.

S. I. Vavilov, P. N. Lebedev Physics Institute, Vestnik Akad. Nauk SSSR (10-11):37, 1937.

1938

V. V. Antonov-Romanovskii, The effect of temperature on the quenching of phosphors, Doklady Akad. Nauk SSSR 20(5):361, 1938; French translation, Compt. rend. acad. sci. URSS 20(5):361, 1938.

N. A. Brilliantov and Z. L. Morgenshtern, The luminescence of x-irradiated rock salt, Zhur. Eksptl. i Teoret. Fiz. 8(4):401, 1938.

M. A. Konstantinova-Shlezinger, The effect of pH on the fluorescence spectrum of a solute, Zhur. Fiz. Khim. 11(5):601, 1938.

M. A. Konstantinova-Shlezinger, A method of fluorescence analysis. A new method of quantitative fluorescence analysis and some cases of its application, Zhur. Fiz. Khim. 11(6):772, 1938.

M. A. Konstantinova-Shlezinger, Determination of ozone in air samples 13 and 14 km above sea level, Doklady Akad. Nauk SSSR 18(6):337, 1938; French translation, Compt. rend. acad. sci. URSS 18(6):337, 1938.

V. L. Levshin and E. L. Rikman, An investigation of the mechanism of phosphorescence of samarium phosphors from the quenching of their luminescence, Doklady Akad. Nauk SSSR 20(6):445, 1938; English translation, Compt. rend. acad. sci. URSS 20(6):445, 1938.

V. L. Levshin, Luminescence of complex molecules, Izvest. Akad. Nauk SSSR, Ser. Fiz. 3(2):337, 1938.

V. L. Levshin, The possibility of interpreting phenomena of polarized luminescence by means of a model of a linear oscillator, in: Trudy FIAN, Vol. 1, No. 4, Moscow-Leningrad, Izd. AN SSSR, p. 19, 1938; English translation, J. Phys. USSR 1(4):265, 1939.

V. L. Levshin and O. A. Pevunova, The nature of luminescence of boron phosphors and the role of boric acid in luminescence, in: Trudy FIAN, Vol. 1, No. 4, Moscow-Leningrad, Izd. AN SSSR, p. 35, 1938.

V. L. Levshin, Cold Light, second edition, revised and supplemented, Moscow-Leningrad, Izd. AN SSSR, 119 pp.

L. A. Tumerman, Dependence of certain properties of fluorescence of solutions on the wavelength of the exciting light, in: Trudy FIAN, Vol. 1, No. 4, Moscow-Leningrad, Izd. AN SSSR, p. 77, 1938.

S. I. Vavilov, The Eye and the Sun. On Light, the Sun, and Vision, third edition, revised and supplemented, Moscow-Leningrad, Izd. AN SSSR, 96 pp., 1938; the same, fourth edition, stereotype, 1941.

S. I. Vavilov, Birefringence (Review), Technical Encyclopedia, second edition, revised and supplemented, Vol. 6, p. 391, 1938.

S. I. Vavilov, Light dispersion, Technical Encyclopedia, second edition, Vol. 6, p. 843, 1938.

S. I. Vavilov, Diffraction, Technical Encyclopedia, second edition, revised and supplemented, Vol. 6, p. 911, 1938.

S. I. Vavilov, Light interference, Technical Encyclopedia, second edition, revised and supplemented, Vol. 9, p. 259, 1938.

S. I. Vavilov, Infrared rays, Technical Encyclopedia, second edition, revised and supplemented, Vol. 9, p. 264, 1938.

S. I. Vavilov, The new physics and dialectical materialism, Pod. znam. marks. (12):27, 1938; also in: S. I. Vavilov, A. A. Maksimov, and V. F. Mitkevich, Materialism and Empiriocriticism of Lenin and Modern Physics, Moscow, Sotsékgiz, p. 66, 1939.

S. I. Vavilov, Optics in the USSR (Review), in: Mathematics and Natural Science in the USSR, Moscow-Leningrad, Izd. AN SSSR, p. 222, 1938.

S. I. Vavilov, Sensitivity of the retina in the ultraviolet spectrum, Doklady Akad. Nauk SSSR 21(8):377, 1938; English translation, Compt. rend. acad. sci. URSS 21(8):373, 1938.

1939

V. N. Alyavdin, V. L. Levshin, and V. V. Fedorov, An investigation of the quenching of luminescence, of some classes of luminescing substances (Al_2O_3Cr, $CdI_2 \cdot MnCl_2$, Zn_2SiO_4-Mn), Doklady Akad. Nauk SSSR 25(2):107, 1939; English translation, Compt. rend. acad. sci. URSS 25(2):106, 1939.

V. V. Antonov-Romanovskii and G. Kochergin, The quenching of alkali phosphors activated by thallium, Doklady Akad. Nauk SSSR 24(5):430, 1939; French translation, Compt. rend. acad. sci. URSS 24(5):430, 1939.

M. A. Konstantinova-Shlezinger, The luminescence properties of crystallophosphors and their chemical structure, Izvest. Akad. Nauk SSSR, Ser. Fiz. 3(1):162, 1939.

M. A. Konstantinova-Shlezinger, Luminescence analysis and its applications, Zavodskaya Lab. 8(7):693, 1939.

M. A. Konstantinova-Shlezinger, Luminescence analysis and its applications, Part 2, Zavodskaya Lab. 8(9):957, 1939.

V. L. Levshin, Photoluminescence, Physics Dictionary, Moscow, ONTI, Vol. 5, p. 491, 1939.

S. I. Vavilov, Science and technology in the epoch of the French revolution, Vestnik Akad. Nauk SSSR (7):15, 1939.

1940

V. N. Alyavdin, V. L. Levshin, and V. V. Fedorov, An investigation of quenching of luminescence of certain classes of luminescing substances, Izvest. Akad. Nauk SSSR, Ser. Fiz. 4(1):173, 1940.

M. A. Konstantinova-Shlezinger, Fluorescence analysis, its uses and theoretical basis, Izvest. Akad. Nauk SSSR, Ser. Fiz. 4(1):114, 1940.

V. L. Levshin and V. N. Tugarinov, The origin of prolonged and brief luminescence of phosphors with organic activators, Doklady Akad. Nauk SSSR 28(2):114, 1940; English translation, Compt. rend. acad. sci. URSS 28(2):115, 1940.

S. I. Vavilov, Academician N. D. Papaleksi, in honor of his sixtieth birthday, Tekhn. Kn. (12):17, 1940.

S. I. Vavilov, Academician V. V. Petrov — an investigator of luminescence, in: Academician V. V. Petrov (1761-1834), Moscow-Leningrad, Izd. AN SSSR, p. 5, 1940.

S. I. Vavilov and A. N. Sevchenko, Quenching of luminescence of rare earth solutions, Doklady Akad. Nauk SSSR 27(6):541, 1940; English translation, Compt. rend. acad. sci. URSS 27(6):541, 1940.

S. I. Vavilov and B. Ya. Sveshnikov, Luminescence analysis in medicine, in: Medical News (information material), 11, Moscow-Leningrad, Medgiz, p. 3, 1940.

S. I. Vavilov, Phosphoroscopic measurements, Doklady Akad. Nauk SSSR 27(2):112, 1940; English translation, Compt. rend. acad. sci. URSS 27(2):115, 1940.

S. I. Vavilov, Polarization of light (Review), Vol. 46, p. 361, 1940.

S. I. Vavilov, The nature of elementary oscillators and the polarization of photoluminescence, Zhur. Eksptl. i Teoret. Fiz. 10(12):1363, 1940; English translation, J. Phys. USSR 3(6):433, 1940.

S. I. Vavilov, The structure of matter, Propaganda i Agitatsiya (23):12, 1940.

L. A. Vinokurov, V. D. Ivanov, and V. L. Levshin, On certain luminescing substances used to increase luminescence and correct the color of mercury light sources, Izvest. Akad. Nauk SSSR, Ser. Fiz. 4(1):134, 1940.

V. A. Yastrebov, The effect of temperature on the light sums and quenching of CaSBi-phosphor, Doklady Akad. Nauk SSSR 28(8):698, 1940; English translation, Compt. rend. acad. sci. URSS 28(8):697, 1940.

1941

V. V. Antonov-Romanovskii, The mechanism of luminescence of phosphors, Doklady Akad. Nauk SSSR 31(9):863, 1941; English translation, Compt. rend. acad. sci. URSS 31(9):863, 1941.

V. V. Antonov-Romanovskii, The mechanism of luminescence of alkali halide phosphors, Izvest. Akad. Nauk SSSR, Ser. Fiz. 5(4-5):523, 1941; English translation, J. Phys. USSR 4(1-2):175, 1941.

V. L. Levshin, Modern investigations into the mechanism of luminescence of semiconductors, Izvest. Akad. Nauk SSSR, Ser. Fiz. 5(4-5):510, 1941.

V. L. Levshin, Investigations into the mechanism of fluorescence of semiconductors (abstract of report to the sixth conference on semiconductors), Zhur. Tekh. Fiz. 11(3):273, 1941.

L. A. Tumerman, The law of quenching of luminescence of complex molecules, Zhur. Eksptl. i Teoret. Fiz. 11(5):515, 1941; English translation, J. Phys. USSR 4(1-2):151, 1941.

L. A. Tumerman, The polarization of fluorescence and the law of its quenching, Doklady Akad. Nauk SSSR 32(7):474, 1941.

S. I. Vavilov, Main trends in modern physics, Priroda (5):3, 1941.

S. I. Vavilov, Luminescent light sources (report to the General Conference of the Academy of Sciences of the USSR, May 30, 1941), Vestnik Akad. Nauk SSSR (7-8):59, 1941.

S. I. Vavilov, Optical methods for the analysis of substances,* in: Collection of Articles to Mark the Twentieth Anniversary of the State Optical Institute (1918-1938), Moscow, Oborongiz, p. 9, 1941; also Trudy GOI, 14:112, 1941.

S. I. Vavilov, The nature of elementary radiators (science news), Zhur. Eksptl. i Teoret. Fiz. 11(1):195, 1941.

S. I. Vavilov and L. A. Tumerman, The nature of light, Nauka i Zhizn (4):29, 1941.

S. I. Vavilov, The development of the idea of matter, Vestnik Akad. Nauk SSSR (1):12, 1941.

1942

V. V. Antonov-Romanovskii, The mechanism of luminescence of alkali halide phosphors, in: Trudy FIAN, Vol. 2, No. 2-3, Moscow-Leningrad, Izd. AN SSSR, p. 157, 1942.

V. V. Antonov-Romanovskii, The luminescence of phosphors during excitation, Doklady Akad. Nauk SSSR 36(4-5):138, 1942.

V. V. Antonov-Romanovskii, The mechanism of luminescence of phosphors, J. Phys. USSR 6(3-4):120, 1942.

E. M. Brumberg, S. I. Vavilov, and Z. M. Sverdlov, Visual measurements of quantum fluctuations.* 1. A comparison of the visual threshold with data of fluctuation measurements, Zhur. Eksptl. i Teoret. Fiz. 12(3-4):93, 1942; English translation, J. Phys. USSR 7(1):1, 1943.

M. A. Konstantinova-Shlezinger, Chemical fluorescence analysis, in: Trudy FIAN, Vol. 2, No. 2-3, Moscow, Izd. AN SSSR, p. 7, 1942.

S. I. Vavilov and T. V. Timofeeva, Visual measurements of quantum fluctuations. 2. Fluctuation during light adaptation of the eye, Zhur. Eksptl. i Teoret. Fiz. 12(3-4):105, 1942; English translation, J. Phys. USSR 7(1):9, 1943.

S. I. Vavilov and T. V. Timofeeva, Visual measurements of quantum fluctuations. 3. Dependence of visual fluctuations on wavelength, Zhur. Eksptl. i Teoret. Fiz. 12(3-4):109, 1942; English translation, J. Phys. USSR 7(1):12, 1943.

S. I. Vavilov, Visual observations of quantum fluctuations of a light field, Izvest. Akad. Nauk SSSR, Ser. Fiz. 6(1-2):74, 1942; English translation, J. Phys. USSR 6(5):224, 1942.

S. I. Vavilov, The theory of concentration quenching of fluorescence of solutions, Doklady Akad. Nauk SSSR 35(4):110, 1942; English translation, Compt. rend. acad. sci. URSS 35(4):100, 1942.

S. I. Vavilov and P. P. Feofilov, Theory of concentration depolarization fluorescence in solutions, Doklady Akad. Nauk SSSR 34(8):243, 1942; English translation, Compt. rend. acad. sci. URSS 34(8):220, 1942.

S. I. Vavilov, Cold light (popular lecture), Moscow-Leningrad, Izd. AN SSSR, p. 31, 1942.

1943

V. V. Antonov-Romanovskii, Increase in phosphorescence during excitation, Doklady Akad. Nauk SSSR 39(8):329, 1943; English translation, Compt. rend. acad. sci. URSS 39(8):299, 1943.

V. V. Antonov-Romanovskii, Screens with an uneven surface, Doklady Akad. Nauk SSSR 41(5):214, 1943; English translation, Compt. rend. acad. sci. URSS 41(5):202, 1943.

V. V. Antonov-Romanovskii, Mechanism of luminescence of phosphors. II. J. Phys. USSR 7(4):153, 1943.

L. A. Tumerman, Infrared rays in the service of industry, Nauka i Zhizn (1-2):39, 1943.

S. I. Vavilov, Isaac Newton, Moscow-Leningrad, Izd. AN SSSR, 216 pp., 1943.

S. I. Vavilov, Isaac Newton (1643-1727), collection of articles in honor of the 300th anniversary of his birth, Academician S. I. Vavilov (ed.), Moscow-Leningrad, Izd. AN SSSR, p. 429, 1943.

S. I. Vavilov, Principles of spectrum transformation of light, Izvest. Akad. Nauk SSSR, Ser. Fiz. 7(1-2):3, 1943.

S. I. Vavilov, Theory of the influence of concentration on the fluorescence of solutions, Zhur. Eksptl. i Teoret. Fiz. 13(1-2):13, 1943; English translation, J. Phys. USSR 7(4):141, 1943.

S. I. Vavilov, Photochemical investigations of Academician P. P. Lazarev (Review), Izvest. Akad. Nauk SSSR, Ser. Fiz. 7(6):193, 1943.

S. I. Vavilov, Ether, light, and matter in the physics of Newton, in: Isaac Newton (1643-1727), collection of articles in honor of the 300th anniversary of his birth, Academician S. I. Vavilov (ed.), Moscow-Leningrad, Izd. AN SSSR, p. 33, 1943.

1944

M. A. Konstantinova-Shlezinger, The fluorescence method of analysis, in: Transactions of the All-Union Conference on Analytical Chemistry of the Academy of Sciences of the USSR, Vol. 3, p. 43, 1944.

S. I. Vavilov, The depolarization of photoluminescence during quenching, Doklady Akad. Nauk SSSR 42(8):344, 1944; English translation, Compt. rend. acad. sci. URSS 42(8):331, 1944.

S. I. Vavilov, Comments on the theory of concentration quenching of fluorescence of solutions, Doklady Akad. Nauk SSSR 45(1):7, 1944; English translation, Compt. rend. acad. sci. URSS 45(1):7, 1944.

S. I. Vavilov, Lenin and modern physics (stenographer's report of a public lecture given in the Hall of Columns in the House of the Unions in Moscow), Moscow, 25 pp., 1944; also in: General Conference of the Academy of Sciences of the USSR, February 14-17, 1944, Moscow-Leningrad Izd. AN SSSR, p. 38, 1944; Vestnik Akad. Nauk SSSR (3):33, 1944; Uspekhi Fiz. Nauk 26(2):113, 1944; Pod. znam. marks. (2-3):36, 1944; Élektrichestvo (4):1, 1944; (5-6):3, 1944.

1945

V. V. Antonov-Romanovskii, The mechanism of luminescence of phosphors (report on the Conference on Problems of Luminescence, Moscow, 1944), Izvest. Akad. Nauk SSSR, Ser. Fiz. 9(4-5):369, 1945.

S. A. Fridman, Investigations into radioluminescence (report to the Conference on Problems of Luminescence, Moscow, October 5-10, 1944), Izvest. Akad. Nauk SSSR, Ser. Fiz. 9(4-5):417, 1945.

S. A. Fridman and A. A. Cherepnev, Luminescing Compositions of Constant and Temporary Action, Moscow, Izd. AN SSSR, 66 pp., 1945.

M A. Konstantinova-Shlezinger and V. S. Krasnova, A quantitative fluorescence method for the determination of traces of oxygen in water, Zavodskaya Lab. 11(6):567, 1945.

M. A. Konstantinova-Shlezinger, Luminescence analysis (report to the Conference on Problems of Luminescence, Moscow, October 5-10, 1944), Izvest. Akad. Nauk SSSR, Ser. Fiz. 9(4-5):469, 1945.

V. L. Levshin, Luminescence of crystalline substances (report to the Conference on Problems of Luminescence, Moscow, October 5-10, 1944), Izvest. Akad. Nauk SSSR, Ser. Fiz. 9(4-5):355, 1945.

V. L. Levshin, President of the Academy of Sciences of the USSR Academician S. I. Vavilov, Izvest. Akad. Nauk SSSR, Ser. Fiz. 9(4-5):269, 1945.

L. A. Tumerman, Quenching of luminescence of complex molecules (report to the Conference on Problems of Luminescence, Moscow, October 5-10, 1944), Izvest. Akad. Nauk SSSR, Ser. Fiz. 9(4-5):328, 1945.

L. A. Tumerman, Problems of illumination (Review), Vestnik Akad. Nauk SSSR (1-2):62, 1945.

S. I. Vavilov, Introductory speech to the Conference on Problems of Luminescence, Moscow, October 5-10, 1944, Izvest. Akad. Nauk SSSR, Ser. Fiz.9(4-5):277, 1945.

S. I. Vavilov, Isaac Newton, second edition, revised and supplemented, Moscow-Leningrad, Izd. AN SSSR, 230 pp., 1945; Rumanian translation, Bucharest, Editura de Stat, 265 pp., 1947; German translation, Vienna, Neue Osterreich, 175 pp., 1948; German translation, Berlin, Akad. Verl. VIII, 214 pp., 1951.

S. I. Vavilov, Lomonosov and Russian Science, Moscow, Mol. Gvardiya, 40 pp., 1945; also, Bol'shevik (6):23, 1945; also, Moscow, Voen. Izd., 46 pp., 1947.

S. I. Vavilov, Luminescent lamps (introductory speech to the Conference on Problems of Luminescence, Moscow, October 5-10, 1944), Izvest. Akad. Nauk SSSR, Ser. Fiz. 9(4-5):487, 1945.

S. I. Vavilov, Comments on the concept and importance of luminescence, Élektrichestvo (1-2):4, 1945.

S. I. Vavilov, A new form of luminescence, Pravda, March 1, 1945, No. 51 (work of scientists of the P. N. Lebedev Physics Institute of the Academy of Sciences of the USSR).

S I. Vavilov, The photoluminescence of solutions (report to the Conference on Problems of Luminescence, Moscow, October 5-10, 1944), Izvest. Akad. Nauk SSSR, Ser. Fiz. 9(4-5):283, 1945.

S. I. Vavilov, Elementary processes of radiation and absorption of light, Priroda (4):9, 1945.

S. I. Vavilov, The development of physics in the Academy of Sciences of the USSR in 220 years, in: Physico-Mathematical Sciences, A. F. Ioffe (ed.), Moscow-Leningrad, Izd. AN SSSR, p. 3, 1945.

S. I. Vavilov, The creative work of the State Optical Institute (25th anniversary of the founding of GOI), Uspekhi Fiz. Nauk 27(1):106, 1945.

S. I. Vavilov, The Physics Cabinet, Physics Laboratory, Physics Institute of the Academy of Sciences of the USSR for 220 Years, Moscow-Leningrad, Izd. AN SSSR, 74 pp., 1945; also Uspekhi Fiz. Nauk 28(1):1, 1946.

S. I. Vavilov, Some remarks on Stokes' law, J. Phys. USSR 9(2):68, 1945.

1946

É. I. Adirovich, The elementary law of quenching corresponding to the band theory of luminescence of crystallophosphors, Doklady Akad. Nauk SSSR 53(4):317, 1946; English translation, Compt. rend. acad. sci. URSS 53(4):313, 1946.

É. I. Adirovich, An elementary law of quenching of crystalline luminescence, Izvest. Akad. Nauk, SSSR, Ser. Fiz. 10(5-6):467, 1946.

M. N. Alentsev, A. F. Belyaev, N. N. Sobolev, and B. M. Stepanov, Optical measurement of the temperature of luminescence of explosion of explosives, Zhur. Eksptl. i Teoret. Fiz. 16(11):990, 1946.

M. N. Alentsev and N. N. Sobolev, The application of the optical method to the determination of the explosion temperature of explosives, Doklady Akad. Nauk SSSR 51(9):691, 1946.

V. V. Antonov-Romanovskii, Recombination phosphorescence, Izvest. Akad. Nauk SSSR,Ser. Fiz. 10(5-6):477, 1946.

V. V. Antonov-Romanovskii, Sm as a flash sensitizer in alkali earth phosphors, Doklady Akad. Nauk SSSR 54(9):779, 1946; English translation, Compt. rend. acad. sci. URSS 54(9):775, 1946.

V. V. Antonov-Romanovskii, V. L. Levshin, Z. L. Morgenshtern, and Z. A. Trapeznikova, Phosphors sensitive to "red light," Doklady Akad. Nauk SSSR 54(1):19, 1946; English translation, Compt. rend. acad. sci. URSS 54(1):19, 1946.

V. L. Levshin, The effect of Mn concentration and temperature on the luminescence of Zn and Mn in ZnS-Mn phosphors, Doklady Akad. Nauk SSSR 54(3):215, 1946; English translation, Compt. rend. acad. sci. URSS 54(2):215, 1946.

V. L. Levshin, The interaction of Zn and Mn luminescence in ZnS-Mn phosphors. The effect of wavelength of the exciting light. 1, Doklady Akad. Nauk SSSR 54(2):127, 1946; English translation, Compt. rend. acad. sci. URSS 54(2):127, 1946.

Z. L. Morgenshtern, Light sum of the flash and phosphorescence. CaS·SrS-Ce, Sm phosphors, Doklady Akad. Nauk SSSR 54(9):791, 1946.

L. A. Tumerman, Luminescent illumination and problems in its introduction, Élektrichestvo (8):3, 1946.

S. I. Vavilov, Resonance migration of excitation energy in fluorescent solutions, Vestnik Leningrad. Univ. (1):5, 1946.

S. I. Vavilov, Soviet Science at a New Stage, Moscow, Izd. AN SSSR, 104 pp., 1946.

S. I. Vavilov, Physics of Lukretsii, in: General Conference of the Academy of Sciences of the USSR, January 16-19, 1946; Reports, Moscow-Leningrad, Izd. AN SSSR, p. 147, 1946; also Izvest. Akad. Nauk SSSR, Ser. Ist. i Filos (1):3, 1946; also Vestnik Akad. Nauk SSSR (2):43, 1946; also Uspekhi Fiz. Nauk 29(1-2): 161, 1946.

S. I. Vavilov, Ether, light and matter in Newtonian physics, in: Moscow University — Isaac Newton Memorial (1643-1943), Moscow, Izd. MGU, 3, 1946.

S. I. Vavilov, Un nouvel aspect de la luminescence, Atomes (1):7, 1946.

S. I. Vavilov, Photoluminescence and thermodynamics (concerning P. Pringsheim's objections to my paper), J. Phys. USSR 10(6):499, 1946.

S. I. Vavilov, Actuelle Probleme der Optik, Mikroskopie 1(3-4):73, 1946.

V. A. Yastrebov, Some features of the luminescence of zinc-cadmium phosphors, Doklady Akad. Nauk SSSR 53(7):609, 1946.

V. A. Yastrebov, The temperature stability of luminescence bands (dissertation for the degree of Candidate of Physical and Mathematical Sciences), in: Trudy FIAN, Vol. 3, No. 2, Moscow-Leningrad, Izd. AN SSSR, p. 121, 1946.

1947

É. I Adirovich, The initial stages of afterglow of crystallophosphors, Doklady Akad. Nauk SSSR 57(2):133, 1947.

É. I. Adirovich, The luminescence of crystallophosphors with constant excitation in the region of absorption of the activator, Doklady Akad. Nauk SSSR 56(6):579, 1947.

É. I. Adirovich, The luminescence of crystallophosphors with constant excitation in the region of the fundamental absorption band, Doklady Akad. Nauk SSSR 57(1):25, 1947.

É. I. Adirovich, A theoretical model of a crystallophosphor and the phenomenon of cold flashing, Doklady Akad. Nauk SSSR 58(9):1926, 1947.

M. N. Alentsev, Curves of darkening of photographic plates at very short exposures, Zhur. Eksptl. i Teoret. Fiz. 17(1):75, 1947.

V. V. Antonov-Romanovskii, Measurement of the absolute yield of the flash of phosphors caused by the action of "red light," Zhur. Eksptl. i Teoret. Fiz. 17(8):708, 1947.

V. V. Antonov-Romanovskii, Special properties of new alkali-earth phosphors sensitive to infrared rays, Doklady Akad. Nauk SSSR 58(5):771, 1947.

A. A. Cherepnev, Zinc sulfide luminophors containing lead, Doklady Akad. Nauk SSSR 56(8):807, 1947.

S. A. Fridman, A. A. Cherepnev, and T. S. Dobrolyubskaya, Afterglow of zinc sulfide luminophors with various activators, Doklady Akad. Nauk SSSR 57(6):563, 1947.

S. A. Fridman, A. A. Cherepnev, and T. S. Dobrolyubskaya, The relationship between zinc and copper luminescence bands in zinc sulfide luminophors, Doklady Akad. Nauk SSSR 57(7):1341, 1947.

S. A. Fridman, A. A. Cherepnev, and T. S. Dobrolyubskaya, Brightness and spectral distribution of luminescence of zinc sulfide luminophors with various activators, Doklady Akad. Nauk SSSR 57(5):451, 1947.

M. D. Galanin, Concentration depolarization of fluorescence during quenching, Doklady Akad. Nauk SSSR, 57(9):883, 1947.

V. L. Levshin, V. V. Antonov-Romanovskii, Z. L. Morgenshtern, and Z. A. Trapeznikova, An investigation of alkali-earth phosphors having high sensitivity to infrared rays, Zhur. Eksptl. i Teoret. Fiz. 17(11):949, 1947.

V. L. Levshin, The interaction of Zn and Mn activators in ZnS-Mn-phosphors, Zhur. Eksptl. i Teoret. Fiz. 17(7):675, 1947.

V. L. Levshin and G. D. Sheremet'ev, The time taken to establish stationary distributions in excited molecules of uranyl salts, Zhur. Eksptl. i Teoret. Fiz. 17(3):209, 1947.

V. L. Levshin, The origin and composition of various forms of luminescence of CaS·SrS-Ce, Sm, La phosphors, Doklady Akad. Nauk SSSR 58(5):779, 1947.

V. L. Levshin, Thirty years of Soviet optics, Nauka i Zhizn (10):35, 1947.

V. L. Levshin, Progress in luminescence analysis, in: Transactions of the Commission for Analytical Chemistry, Vol. I(IV), p. 128 (1947).

Z. L. Morgenshtern, Accumulation of the light sum in phosphors sensitive to infrared rays, Doklady Akad. Nauk SSSR 58(5):783, 1947.

N. D. Papaleksi (ed.), Course of Physics, Vol. 2, Electricity; Optics; Nuclear Physics; Compiled by Professors S. M. Rytov, V. L. Levshin, E. L. Feinberg, L. V. Groshev, Moscow-Leningrad, Gostekhizdat, 1947.

Z. A. Trapeznikova, Direct proof of the luminescent action of exciting light for SrS-Eu, Sm phosphors, Doklady Akad. Nauk SSSR 58(5):791, 1947.

L. A. Tumerman, A photoelectric method for the measurement of the degree of polarization of radiation, Doklady Akad. Nauk SSSR 58(9):1945, 1947.

L. A. Tumerman, Experimental methods for the investigation of fast relaxation processes, Uspekhi Fiz. Nauk 33(2):218, 1947.

S. I. Vavilov, Luminescence and its duration, in: Academy of Sciences of the USSR, Jubilee Review in Honor of the Thirtieth Anniversary of the Great October Socialist Revolution, Part 1, Moscow-Leningrad, Izd. AN SSSR, p. 377, 1947.

S. I. Vavilov, Luminescence and its uses in the technology of light, Elektrichestvo (12):3, 1947.

S. I. Vavilov, The optical work and views of M. V. Lomonosov, in: B. M. Menshutkin, Biography of Mikhail Vasil'evich Lomonosov; third edition with supplement by P. N. Berkov, S. I. Vavilov, and L. B. Modzalevskii, Moscow-Leningrad, Izd. AN SSSR, p. 147, 1947.

S. I. Vavilov, Newton and the atomic theory, Royal Society Newton Tercentenary celebrations, July 15-19, 1946, Cambridge University Press, p. 43, 1947.

V. A. Yastrebov, The temperature stability of luminescence bands, Zhur. Eksptl. i Teoret. Fiz. 17(2):140, 1947.

1948

É. I. Adirovich, Radiationless electron transitions in a disturbed crystal structure, Doklady Akad. Nauk SSSR 63(6):635, 1948.

É. I. Adirovich, A reduced law of quenching of phosphorescence in crystals, Zhur. Eksptl. i Teoret. Fiz. 18(1):58, 1948.

É. I. Adirovich, The kinetics of successive reactions, Doklady Akad. Nauk SSSR 61(3):467, 1948.

É. I. Adirovich, Kinetics of afterglow of crystallophosphors, Doklady Akad. Nauk SSSR 60(3):361, 1948.

É. I. Adirovich, Electron states in a disturbed crystal structure, Doklady Akad. Nauk SSSR 63(2):111, 1948.

M. N. Alentsev, Dependence of the yield of fluorescence of iodine vapors on the wavelength of the exciting light, Doklady Akad. Nauk SSSR 62(5):607, 1948.

V. V. Antonov-Romanovskii, The luminescing action of exciting light (report to the Second All-Union Conference on Luminescence), Uspekhi Fiz. Nauk 36(4):561, 1948.

A. A. Cherepnev, Photoluminescence of zinc sulfides containing tin, Doklady Akad. Nauk SSSR 62(6):767, 1948.

A. A. Cherepnev and T. S. Dobrolyubskaya, Zinc sulfide luminophors containing cobalt, Doklady Akad. Nauk SSSR 62(3):325, 1948.

S. A. Fridman and A. A. Cherepnev, A new type of zinc sulfide luminophors, Doklady Akad. Nauk SSSR 59(1):53, 1948.

S. A. Fridman and N. O. Chechik, Photoelectric photometry of light compositions of temporary action, Zhur. Tekhn. Fiz. 18(1):35, 1948.

M. D. Galanin, The duration of the initial luminescence of phosphors, Doklady Akad. Nauk SSSR 60(5):783, 1948.

M. A. Konstantinova-Shlezinger and N. A. Gorbacheva, The theory of chromatographic analysis, Zhur. Anal. Khim. 3(4):213, 1948.

M. A. Konstantinova-Shlezinger, Luminescence Analysis, Moscow-Leningrad, Izd. AN SSSR, 515 pp., 1948.

V. L. Levshin, The effect of electron distribution over localization levels on the course of various luminescence processes for CaS·SrS-Ce, Sm, La phosphors and the number of repeated localizations of electrons, Zhur. Eksptl. i Teoret. Fiz. 18(2):149, 1948.

V. L. Levshin, The nature of various forms of luminescence of phosphors with deep localization levels, Zhur. Eksptl. i Teoret. Fiz. 18(1):82, 1948.

V. L. Levshin, Various processes of luminescence of crystallophosphors, Izvest. Akad. Nauk SSSR, Ser. Fiz. 12(3):217, 1948.

S. I. Vavilov, Luminescence and its duration, in: General Conference of the Academy of Sciences of the USSR in Honor of the Thirtieth Anniversary of the Great October Socialist Revolution, Reports, Moscow-Leningrad, Izd. AN SSSR, p. 432, 1948.

S. I. Vavilov, Mikhail Vasil'evich Lomonosov (1711-1765), in: Men of Russian Science, Vol. I, Moscow-Leningrad, Gostekhizdat, p. 63, 1948.

S. I. Vavilov, A few words on luminescent lamps, in: R. A. Nilender, Luminescent Lamps and Their Uses, Moscow-Leningrad, Gosénergoizdat, p. 3, 1948.

S. I. Vavilov, Petr Nikolaevich Lebedev (1866-1912), in: Men of Russian Science, Vol. I, Moscow-Leningrad, Gostekhizdat, p. 241, 1948.

S. I. Vavilov, Soviet science in the service of the Motherland, in: Men of Russian Science, Vol. I, Moscow-Leningrad, Gostekhizdat, p. 21, 1948.

S. I. Vavilov, Lessons of the past and possibilities for the study of luminescence, Priroda (12):11, 1948.

S. I. Vavilov, Experimental investigations of light quanta fluctuations by the visual method, Uspekhi Fiz. Nauk 36(3):247, 1948.

1949

É. I. Adirovich, The zone theory of crystals and the phenomenon of the cold flash, Izvest. Akad. Nauk SSSR, Ser. Fiz. 13(1):101, 1949.

É. I. Adirovich, The effect of local electron states on the permittivity of crystallophosphors, Doklady Akad. Nauk SSSR 66(4):601, 1949.

É. I. Adirovich, The connection between the probabilities of direct and inverse spectral transformation of light in the process of luminescence, Doklady Akad. Nauk SSSR 69(6):759, 1949.

M. N. Alentsev, Dependence of the yield of luminescence of crystallophosphors on the wavelength of the exciting light, Doklady Akad. Nauk SSSR 64(4):479, 1949.

L. I. Anikina and V. V. Antonov-Romanovskii, An investigation of the change in absorption during the excitation of phosphors, Doklady Akad. Nauk SSSR 68(4):669, 1949.

V. V. Antonov-Romanovskii, The luminescing action of exciting light in phosphors (report to the Second All-Union Conference, Moscow, 1948), Izvest. Akad. Nauk SSSR, Ser. Fiz. 13(1):91, 1949.

V. V. Antonov-Romanovskii and M. I. Épshtein, Measurement of the absolute yield of luminescence of powdered phosphors, Doklady Akad. Nauk SSSR 64(4):483, 1949.

V. V. Antonov-Romanovskii, The temperature luminescence of phosphors, Doklady Akad. Nauk SSSR 68(3):457, 1949.

V. V. Antonov-Romanovskii, V. L. Levshin, Z. L. Morgenshtern, and Z. A. Trapeznikova, The mechanism of the flash of SrS phosphors activated by rare earth activators and the interaction of activators (report to the Second All-Union Conference on Luminescence, Moscow, 1948), Izvest. Akad. Nauk SSSR, Ser. Fiz. 13(1):75, 1949.

V. V. Antonov-Romanovskii and E. S. Krylova, The analysis of quenching curves of powdered phosphors, Zhur. Eksptl. i Teoret. Fiz. 19(1):63, 1949.

A. A. Cherepnev and T. S. Dobrolyubskaya, The problem of the formation of luminescence centers of ZnS-Cu luminophors, Doklady Akad. Nauk SSSR 66(4):621, 1949.

A. V. Karyakin and M. D. Galanin, Duration of the excited state of molecules of anthraquinone derivatives in vapors and adsorbates, Doklady Akad. Nauk SSSR 66(1):37, 1949.

M. A. Konstantinova-Shlezinger, Luminescence properties of crystallophosphors and their chemical structure (report to the Second All-Union Conference on Luminescence, Moscow, 1948), Izvest. Akad. Nauk SSSR, Ser. Fiz. 13(2):237, 1949.

V. L. Levshin, Cold Luminescence, Moscow, Izd. Pravda, 39 pp., 1949.

S. I. Vavilov, Academy of Sciences of the USSR, Bol'shaya Sov'etskaya Éntsiklopediya, second edition, Vol. I, p. 570, 1949.

S. I. Vavilov, Academy of Sciences of the USSR and the Development of Soviet Science (report to the General Assembly of the Academy of Sciences of the USSR, January 7, 1949), Vestnik Akad. Nauk SSSR (2):38, 1949.

S. I. Vavilov, Opening speech to the Second Conference on Problems of Luminescence and the Use of Light Composition, Moscow, 1948, Izvest. Akad. Nauk SSSR, Ser. Fiz. 13(1):5, 1949.

S. I. Vavilov, Lomonosov's Law (Law of Preservation and Indestructability of Matter and Motion), Pravda, January 5, 1949; also Fiz. v Shkole (5):6, 1951.

S. I. Vavilov and M. D. Galanin, Radiation and absorption of light in a system of inductively bound molecules, Doklady Akad. Nauk SSSR 67(5):811, 1949.

S. I. Vavilov, Lenin and modern physics, in: Modern Problems of Science and Technology, Moscow, p. 5, 1949.

S. I. Vavilov, Lenin and philosophical problems of modern physics, Uspekhi Fiz. Nauk 38(2):145, 1949.

S. I. Vavilov, "Materialism and empiriocriticism" of V. I. Lenin and philosophical problems of modern physics, Vestnik Akad. Nauk SSSR (6):30, 1949.

S. I. Vavilov, On "hot" and "cold" light (thermal radiation and luminescence), Moscow-Leningrad, Izd. AN SSSR, 75 pp., 1949.

S. I. Vavilov, Soviet science in the service of the Motherland, in: Science and Life, Collection, Moscow, p. 3, 1949.

S. I. Vavilov, M. D. Galanin, and F. M. Pekerman, Experimental investigations of the migration of energy in fluorescing solutions (report to the Second Conference on Luminescence, Moscow, 1948), Izvest. Akad. Nauk SSSR, Ser. Fiz. 13(1):18, 1949; German translation, Abhandlungen aus der sowjetischen Physik, Folge I, Berlin, Verl. Kultur und Fortschritt, p. 9, 1951.

N. D. Zhevandrov, V. L. Levshin, and K. K. Mozgova, The effect of molecular structure of 9, 10-diaryldiamino-anthracenes on their optical properties (report to the Second All-Union Conference on Luminescence, Moscow, 1948), Izvest. Akad. Nauk SSSR, Ser. Fiz. 13(1):49, 1949.

1950

É. I. Adirovich, Luminescence and laws of spectral transformation of light, Uspekhi Fiz. Nauk 40(3):341, 1950.

É. I. Adirovich, The anomalously long duration of some dipole radiations, Izvest. Akad. Nauk SSSR, Ser. Fiz. 14(3):366, 1950.

É. I. Adirovich, Elementary law of quenching of crystallophosphors and the phenomenon of the cold flash (dissertation for the Doctorate of Physical and Mathematical Sciences), in: Trudy FIAN, Vol. 5, Moscow-Leningrad, Izd. AN SSSR, p. 387, 1950.

M. N. Alentsev, Dependence of the yield of photoluminescence on the wavelength of the exciting light (dissertation for the degree of Candidate of Physical and Mathematical Sciences), in: Trudy FIAN, Vol. 5, Moscow-Leningrad, Izd. AN SSSR, p. 499, 1950.

L. I. Anikina and V. V. Antonov-Romanovskii, The influence of unevenness of excitation on the flash properties of phosphors sensitive to infrared rays, Doklady Akad. Nauk SSSR 71(4):637, 1950.

V. V. Antonov-Romanovskii and I. P. Shchukin, Measurement of absorption of ZnS-Cu, Co phosphor, Doklady Akad. Nauk SSSR 71(3):445, 1950.

M. D. Galanin, Time of the excited state of molecules and the fluorescence properties of solutions (dissertation for the degree of Candidate of Physical and Mathematical Sciences), in: Trudy FIAN, Vol. 5, Moscow-Leningrad, Izd. AN SSSR, p. 339, 1950.

M. D. Galanin, Measurement of the duration of fluorescence with the "phase fluorometer," Doklady Akad. Nauk SSSR 73(5):925, 1950.

M. L. Galanin, The effect of temperature on the duration of luminescence of fluorescein solutions, Doklady Akad. Nauk SSSR 70(6):989, 1950.

V. L. Levshin, Bioluminescence, Bol'shaya Sov'etskaya Éntsiklopediya, second edition, Vol. 5, p. 210, 1950.

V. L. Levshin and L. V. Veits, Change in the optical properties of ZnS-Cu, ZnS-Mn and ZnS, CdS-Mn phosphors during their mechanical size reduction, * Zhur. Eksptl. i Teoret. Fiz. 20(5):411, 1950.

Z. L. Morgenshtern, The use of flash phosphors for photography in the infrared region of the spectrum, Doklady Akad. Nauk SSSR 74(3):493, 1950.

Z. A. Trapeznikova, The interaction of activators in phosphors, Doklady Akad. Nauk SSSR 74(3):465, 1950.

S. I Vavilov, The 'Eye and the Sun' (Light, Sun, and Vision), Fifth edition, revised and supplemented, Moscow-Leningrad, Izd. AN SSSR, 122 pp., 1950; Ukrainian translation, Kiev, Rad. Shkola, 124 pp., 1953; Polish translation, Warsaw, Ksiazka i Wiedza, 145 pp., 1952; Hungarian translation, Bucharest, ARLUS— Cartea Rusa, 141 pp., 1953; Georgian translation, Tbilisi, AN Graz.SSR, 196 pp., 1953; German translation, Berlin, Akad. Verl., 1953; Estonian translation, Tallinn, Éstgosizdat, 123 pp., 1954; English translation, Moscow, Foreign Languages Publishing House, 136 pp., 1955; French translation, Moscow, Ed. en lang. étr., 143 pp., 1955.

S. I. Vavilov, Notes on the depolarization of photoluminescence during quenching, Doklady Akad. Nauk SSSR 73(6):1145, 1950.

S. I. Vavilov, The Microstructure of Light (Investigation and Essays), Moscow, Izd. AN SSSR, 198 pp., 1950; Rumanian translation, Bucharest, 187 pp., 1953; Polish translation, Warsaw, 227 pp., 1953; German translation, Berlin, Akad. Verl., VII, 161, 1954; Hungarian translation, Budapest, Akad. kiado, 175 pp., 1955; Ukrainian translation, Kiev, Rad.Shkola, 175 pp., 1956.

N. D. Zhevandrov, Polarization spectra of anthracene derivatives, Doklady Akad. Nauk SSSR 74(1):25, 1950.

1951

É. I. Adirovich, Luminescence and photoconductivity of crystallophosphors under conditions of weak excitation, Zhur. Eksptl. i Teoret. Fiz. 21(2):275, 1951.

É. I. Adirovich, Mechanism of ionization and temperature dependence of photoconductivity and luminescence of crystals, Doklady Akad. Nauk SSSR 76(5):665, 1951.

M. N. Alentsev and L. A. Vinokurov, Measurement of the absolute yield of ultraviolet luminescence (report to the Third Conference on Luminescence and the Use of Light Compositions, Moscow, 1951), Izvest. Akad. Nauk SSSR, Ser. Fiz. 15(6):725, 1951.

M. N. Alentsev, Calorimetric measurement of the yield of fluorescence, Zhur. Eksptl. i Teoret. Fiz. 21(2):133, 1951.

M. N. Alentsev, S. M. Bukshtein, I. I. Kalinichenko, T. V. Kuzina, F. M. Pekerman, and A. V. Chistyakova, Luminophors for luminescent lamps (report to the Third Conference on Luminescence and the Use of Light Compositions, Moscow, 1951), Izvest. Akad. Nauk SSSR, Ser. Fiz. 15(6):824, 1951.

L. I. Anikina, The fluorescence and phosphorescence of alkali-earth phosphors containing Ce activators, Zhur. Eksptl. i Teoret. Fiz. 21(2):310, 1951.

V. V. Antonov-Romanovskii, The flashing and quenching action of exciting light on crystallophosphors (report to the Third Conference on Luminescence and the Use of Light Compositions, Moscow, 1951), Izvest. Akad. Nauk SSSR, Ser. Fiz. 15(5):637, 1951.

V. V. Antonov-Romanovskii, Features of the γ-luminescence of phosphors, Zhur. Eksptl. i Teoret. Fiz. 21(2): 269, 1951.

A. A. Cherepnev, The dispersion state of activators in luminophors (report to the Third Conference on Luminescence and the Use of Light Compositions, Moscow, 1951), Izvest. Akad. Nauk SSSR, Ser. Fiz. 15(6):742, 1951.

A. A. Cherepnev, The activation of zinc sulfide luminophors with copper, Zhur. Eksptl. i Teoret. Fiz. 21(2):322, 1951.

M. D. Galanin, Quenching by absorbing substances and sensitized fluorescence in solutions (report to the Third Conference on Luminescence and the Use of Light Compositions, Moscow, 1951), Izvest. Akad. Nauk SSSR, Ser. Fiz. 15(5):543, 1951.

M. D. Galanin and L. V. Levshin, Quenching of fluorescence of solutions by absorbing substances. I, Zhur. Eksptl. i Teoret. Fiz. 21(2):121, 1951.

M. D. Galanin, Quenching of fluorescence of solutions by absorbing substances. II, Zhur. Eksptl. i Teoret. Fiz. 21(2):126, 1951.

M. D. Galanin and I. M. Frank, Quenching of fluorescence by a medium which absorbs light, Zhur. Eksptl. i Teoret. Fiz. 21(2):114, '1951.

N. A. Gorbacheva, Features of the luminescence of $Cd_2P_2O_7$-Mn, Pb phosphor, Zhur. Eksptl. i Teoret. Fiz. 21(2):305, 1951.

M. A. Konstantinova-Shlezinger, The nature of luminescence centers, Zhur. Eksptl. i Teoret. Fiz. 21(2):252, 1951.

M. A. Konstantinova-Shlezinger, Review of work on luminescence analysis published during the last three years (report to the Third Conference on Luminescence and Use of Light Compositions, Moscow, 1951), Izvest. Akad. Nauk SSSR, Ser. Fiz. 15(6):762, 1951.

M. A. Konstantinova-Shlezinger, Collection of Abstracts on Luminescence Analysis, Moscow, Izd. AN SSSR, 64 pp., 1951.

V. L. Levshin, S. I. Vavilov — founder and head of the Soviet school of luminescence (report to the Third Conference on Luminescence and Use of Light Compositions, Moscow, 1951), Izvest. Akad. Nauk SSSR, Ser. Fiz. 15(5):513, 1951.

V. L. Levshin and T. M. Tarasova, Effect of molecular structure and temperature of the medium on the luminescence and absorption of complex molecules (report to the Third Conference on Luminescence and Use of Light Compositions, Moscow, 1951), Izvest. Akad. Nauk SSSR, Ser. Fiz. 15(5):573, 1951.

V. L. Levshin, Luminescence of activated crystals, Uspekhi Fiz. Nauk 43(3):426, 1951; also in collection: Problems of Physical Optics and Other Problems of Physics (collection of articles to honor the memory of S. I. Vavilov), Moscow-Leningrad, GITTL, p. 76, 1951.

V. L. Levshin, Photoluminescence of Solids and Liquids, Moscow-Leningrad, GITTL, 456 pp., 1951; Hungarian translation, Budapest, Akad. kiado, 552 pp., 1956; Chinese translation, 407 + 15 pp., 1958.

Z. L. Morgenshtern, The mechanism of luminescence of diamonds, Zhur. Eksptl. i Teoret. Fiz. 21(2):230, 1951.

E. G. Teremetskaya and Z. A. Trapeznikova, Centers of luminescence and factors affecting the processes for the preparation of crystallophosphors (report to the Third Conference on Luminescence and the Use of Light Composition, Moscow, 1951), Izvest. Akad. Nauk SSSR, Ser. Fiz. 15(6):730, 1951.

Z. A. Trapeznikova, The interaction of activators in phosphors, Zhur. Eksptl. i Teoret. Fiz. 21(2):283, 1951.

L. A. Vinokurov, V. L. Levshin, and E. G. Baranova, An investigation of the luminescence of zinc sulfide phosphors, Zhur. Eksptl. i Teoret. Fiz. 21(2):236, 1951.

L. A. Vinokurov, Photocopying using a luminescing screen, Zhur. Eksptl. i Teoret. Fiz. 21(2):338, 1951.

V. A. Yastrebov, Nonexponential quenching of solid aromatic hydrocarbons, Zhur. Eksptl. i Teoret. Fiz. 21(2):164, 1951.

1952

É. I. Adirovich, Comments of M. N. Adamov relative to the conditions of temperature quenching of luminescence of crystallophosphors proposed by É. I. Adirovich, Zhur. Eksptl. i Teoret. Fiz. 22(2):246, 1952.

V. V. Antonov-Romanovskii, The kinetics of phosphorescence, Doklady Akad. Nauk SSSR 85(3):517, 1952.

M. A. Konstantinova-Shlezinger, N. A. Gorbacheva, and E. I. Panasyuk, The characteristics of a class of photoluminophors based on sulfates, Zhur. Eksptl. i Teoret. Fiz. 23(5):588, 1952.

V. L. Levshin, S. I. Vavilov — teacher of scientific cadres, in: In Memory of Sergei Ivanovich Vavilov, Moscow, Izd. AN SSSR, p. 89, 1952.

I. G. Petrovskii and V. L. Levshin, The work of S. I. Vavilov in the section of physical and mathematical sciences of the Academy of Sciences of the USSR, in: In Memory of Sergei Ivanovich Vavilov, Moscow, Izd. AN SSSR, p. 17, 1952.

K. V. Shalimova and T. P. Belikova, Duration of the excited state of certain phosphors, Doklady Akad. Nauk SSSR 82(5):713, 1952.

S. I. Vavilov, Galileo, Bol'shaya Sov'etskaya Éntsiklopediya, second edition, Vol. 10, p. 125, 1952.

S. I. Vavilov, Birefringence, Bol'shaya Sov'etskaya Éntsiklopediya, second edition, Vol. 13, p. 474, 1952.

L. A. Vinokurov, The temperature and infrared quenching of ZnS-Co, Co phosphor, Doklady Akad. Nauk SSSR 85(3):529, 1952.

N. D. Zhevandrov, Polarization of fluorescence of organic crystals, Doklady Akad. Nauk SSSR 83(5):677, 1952.

1953

L. I. Anikina, The effect of the luminescing action of exciting light on the luminescence yield, Doklady Akad. Nauk SSSR 88(1):41, 1953.

V. V. Antonov-Romanovskii, E. E. Bukki, and L. A. Vinokurov, Absorption (in spectrum) of zinc sulfide phosphors activated by cobalt and nickel, Zhur. Eksptl. i Teoret. Fiz. 25(6):745, 1953.

M. V. Fok, Wide-angle interference from a quadrupole light source, Doklady Akad. Nauk SSSR 89(3):439, 1953.

V. L. Levshin, Infrared rays, Bol'shaya Sov'etskaya Éntsiklopediya, second edition, Vol. 18, p. 332, 1953; also, Physics Dictionary, ONTI, Vol. 2, p. 706, 1937.

V. L. Levshin, Scientific and teaching activity of S. I. Vavilov, Vestnik Moskov. Univ., Ser. Fiz.-Mat. i Estestven. Nauk 5(3):3, 1953.

Z. L. Morgenshtern, Dependence of some luminescence properties of SrS-Ce, Sm and ScS-Eu, Sm phosphors on the wavelength of the exciting light, Zhur. Eksptl. i Teoret. Fiz. 25(4):491, 1953.

V. F. Tunitskaya, The origin of separate luminescence bands of CaS-Bi phosphors, Doklady Akad. Nauk SSSR 91(3):507, 1953.

V. A. Yastrebov, The law of quenching of luminescence of solid organic substances, Doklady Akad. Nauk SSSR 90(6):1015, 1953.

1954

M. N. Alentsev, V. V. Antonov-Romanovskii, and L. A. Vinokurov, Dependence of the yield of green luminescence of ZnS-Cu phosphor on the intensity of excitation, Doklady Akad. Nauk SSSR 96(6):1133, 1954.

M. N. Alentsev and A. A. Cherepnev, Dependence of the absorption and luminescence spectra of ZnS-Cu phosphor on the copper concentration, Zhur. Eksptl. i Teoret. Fiz. 26(4):473, 1954.

V. V. Antonov-Romanovskii, Determination of the absorption coefficient of powdered phosphors, Zhur. Eksptl. i Teoret. Fiz. 26(4):459, 1954.

L. M. Belyaev, M. D. Galanin, Z. L. Morgenshtern, and Z. A. Chizhikova, Dependence of the yield of γ-photoluminescence of KI-Tl crystals on the thallium concentration, Doklady Akad. Nauk SSSR 99(5):691, 1954.

M. D. Galanin and Z. A. Chizhikova, Yield of photoluminescence of some organic crystals, Zhur. Eksptl. i Teoret. Fiz. 26(5):624, 1954.

V. I. Gribkov and N. D. Zhevandrov, An investigation of polarization characteristics of the luminescence of complex organic molecules by a photoelectrical method, Doklady Akad. Nauk SSSR 98(4):565, 1954.

M. A. Konstantinova-Shlezinger, E. G. Vasil'eva, and Z. N. Repukhova, A magnesium-lithium-tungstate crystallophosphor with manganese activator, Doklady Akad. Nauk SSSR 95(2):241, 1954.

M. A. Konstantinova-Shlezinger, Collection of abstracts on luminescence analysis, Moscow, Izd. AN SSSR, 104 pp., 1954.

V. L. Levshin, Sergei Ivanovich Vavilov (essay on his life and work), in: S. I. Vavilov, Collected Works, Vol. 1, Moscow, Izd. AN SSSR, p. 7, 1954.

V. L. Levshin, Luminescence, Bol'shaya Sov'etskaya Éntsiklopediya, 1954; also Physics Dictionary, ONTI, Vol. 3, p. 377, 1937.

1955

M. N. Alentsev, V. V. Antonov-Romanovskii, B. I. Stepanov, and M. V. Fok, The yield of resonance fluorescence of atoms, Zhur. Eksptl. i Teoret. Fiz. 28(2):253, 1955.

V. V. Antonov-Romanovskii, B. I. Stepanov, M. V. Fok, and A. P. Khapalyuk, Luminescence yield of a system with three energy levels, Doklady Akad. Nauk SSSR; V. V. Antonov-Romanovskii, I. B. Keirim-Markus, M. S. Poroshina, and Z. A. Trapeznikova, Dosimetry of radioactive radiation using flashing phosphors, in: Session of the Academy of Sciences of the USSR on the Peaceful Uses of Atomic Energy, July 1-5, 1955; session of the physical and mathematical sciences section, Moscow, Izd. AN SSSR, p. 342, 1955.

V. V. Antonov-Romanovskii and L. A. Vinokurov, Reduction in the luminescence yield of phosphors during intensive excitation, Zhur. Eksptl. i Teoret. Fiz. 29(6):830, 1955; English translation, Soviet Phys. JETP 2(4):711, 1956.

V. V. Antonov-Romanovskii, Determination of the absorption coefficient of powdered phosphors, Magyar Fizikai Folyóirat 3(1):87, 1955.

T. P. Belikova, Phosphorescence of ZnS-Cu crystallophosphor during excitation by an electron beam, Zhur. Eksptl. i Teoret. Fiz. 29(6):905, 1955.

L. M. Belyaev, M. D. Galanin, Z. L. Morgenshtern, and Z. A. Chizhikova, Dependence of the yield of γ- and photoluminescence of NaI-Tl crystals on the thallium concentration, Doklady Akad. Nauk SSSR 105(1):57, 1955.

E. E. Bukke, Change in the permittivity of phosphors under the action of infrared light, Zhur. Eksptl. i Teoret. Fiz. 28(4):507, 1955.

A. A. Cherepnev, The state of the copper activator in zinc sulfide luminophors, Zhur. Eksptl. i Teoret. Fiz. 28(4):458, 1955.

M. D. Galanin, The effect of concentration on the luminescence of solutions, Zhur. Eksptl. i Teoret. Fiz. 28(4):485, 1955.

M. D. Galanin, Reabsorption of luminescence in thin layers, Doklady Akad. Nauk SSSR 105(4):700, 1955.

Hsü Hsü-yung, Determination of the effective cross sections of captive and recombination of thermal electrons in ZnS-Cu, Co phosphor, Doklady Akad. Nauk SSSR 103(4):585, 1955.

V. L. Levshin, Luminescent illumination and ways for its introduction, Vestnik Akad. Nauk SSSR (10):54, 1955.

V. L. Levshin and A. G. Laktionov, Absorption of complex molecules which are in the metastable state, Doklady Akad. Nauk SSSR 103(1):61, 1955.

V. L. Levshin, Luminescence of molecules and crystals, Nauka i Zhizn (5):17, 1955.

V. L. Levshin, Transformation and transfer of energy in a substance during photoluminescence, Magyar Fizikai Folyóirat (2):529, 1955.

Z. L. Morgenshtern, Measurement of the absolute quantum yield of photoluminescence of alkali halide crystals, Zhur. Eksptl. i Teoret. Fiz. 29(6):903, 1955.

Z. L. Morgenshtern, Spectral dependence of the yield of photoluminescence of alkali iodides activated with thallium, Doklady Akad. Nauk SSSR 105(2):250, 1955.

I. P. Shchukin, Some problems in the recombination luminescence of KCl-Tl phosphor during various forms of excitation, Zhur. Eksptl. i Teoret. Fiz. 29(6):834, 1955.

I. P. Shchukin, The ratio of effective cross sections of recombination and capture of electrons and the concentration of ionic vacancies in a KCl-Tl crystal, Doklady Akad. Nauk SSSR 104(2):211, 1955.

S. I. Vavilov, Polarization of light, Bol'shaya Sov'etskaya Éntsiklopediya, second edition, Vol. 34, p. 114, 1955.

N. D. Zhevandrov, The effect of migration of energy on the polarization of fluorescence of single crystals, Doklady Akad. Nauk SSSR 100(3):455, 1955.

N. D. Zhevandrov, Relationship between the polarization of luminescence and other optical properties of anthracene derivatives and their structure, Trudy FIAN, Vol. 6, Moscow-Leningrad, Izd. AN SSSR, p. 121, 1955.

1956

É. I. Adirovich, Some Problems in the Theory of Luminescence of Crystals, Moscow-Leningrad, Gostekhizdat, 350 pp. Also 1956; German translation, Berlin, Akad. Verlag, 298 pp., 1953.

É. I. Adirovich, La formule de Becquerel et la loi élémentaire du déclin de la luminescence des phosphores cristallins, J. phys. rad. 17(8-9):705, 1956.

É. I. Adirovich, La formule de Becquerel et la loi élémentaire du déclin de la luminescence des phosphores cristallins (Colloques internationaux du Centre National de la Recherche Scientifique LXXII), La luminescence des corps cristallins anorganiques, Paris, May 21-26, 1956, p. 97.

M. N. Alentsev, Dependence of the luminescence yield of ZnS-Cu phosphor on the wavelength of the exciting light, Optika i Spektroskopiya 1(2):240, 1956.

L. I. Anikina, Effect of the luminescing action of exciting light on the photoluminescence yield of crystallophosphors (dissertation for the degree of candidate of physical and mathematical sciences), in: Trudy FIAN, Vol. 7, Moscow, Izd. AN SSSR, p. 5, 1956.

V. V. Antonov-Romanovskii and L. A. Vinokurov, Kinetics of the luminescence of ZnS-Cu, Co phosphor in the region of independence of the luminescence yield on the intensity of the exciting light, Optika i Spektroskopiya 1(1):71, 1956.

V. V. Antonov-Romanovskii and Z. A. Trapeznikova, Two recombination mechanisms of the luminescence of the Sm^{+++} ion in ZnS-Sm, Cu (NaCl) phosphor, Optika i Spektroskopiya 1(2):204, 1956.

V. V. Antonov-Romanovskii and L. A. Vinokurov, Nature of the luminescence losses of ZnS-Cu, Co phosphor in the region of independence of the yield on the intensity of the exciting light, Optika i Spektroskopiya 1(1):66, 1956.

V. V. Antonov-Romanovskii, Sur le rendement de la luminescence dans les phosphores cristallins, J. phys. rad. 17(8-9):694, 1956.

T. P. Belikova, M. D. Galanin, and Z. A. Chizhikova, Transfer of excitation energy from the solvent to the dissolved luminophor in liquid and solid solutions, Izvest. Akad. Nauk SSSR, Ser. Fiz. 20(4):384, 1956.

T. P. Belikova and M. D. Galanin, Sensitization of photoluminescence by the solvent, Optika i Spektroskopiya 1(2):168, 1956.

A. A. Cherepnev, Zinc oxide in a zinc sulfide luminophor, Optika i Spektroskopiya 1(2):272, 1956.

Z. A. Chizhikova and M. D. Galanin, Yields of γ- and photoluminescence of organic crystals, Zhur. Eksptl. i Teoret. Fiz. 30(1):187, 1956.

L. D. Derkacheva, Concentration effects in solutions of cyanine dyes, Izvest. Akad. Nauk SSSR, Ser. Fiz. 20(4):410, 1956.

M. D. Galanin and A. P. Grishin, The absolute yield of luminescence during γ-scintillations in a crystal of naphthalene with anthracene, Zhur. Eksptl. i Teoret. Fiz. 30(1):33, 1956.

M. D. Galanin, The luminescence of organic substances under the action of particles and hard radiation, Izvest. Akad. Nauk SSSR, Ser. Fiz. 20(4):392, 1956.

M. D. Galanin and Z. A. Chizhikova, Transfer of excitation energy in crystals of anthracene with an impurity of naphthacene, Optika i Spektroskopiya 1(2):175, 1956.

Hsü Hsü-yung, Determination of the ratio of effective cross sections of capture and recombination of optical electrons in ZnS-Cu, Co crystallophosphor, Doklady Akad. Nauk SSSR 106(5):818, 1956.

Hsü Hsü-yung, Determination of the effective cross sections of captive and recombination of thermal electrons in ZnS-Cu, Co phosphor, Optika i Spektroskopiya 1(2):264, 1956.

M. A. Konstantinova-Shlezinger, Academician S. I. Vavilov and his role in the development of luminescence analysis, Zhur. Anal. Khim. 11(1):115, 1956.

V. L. Levshin, The effect of association and other physicochemical factors on luminescence and the absorption of complex molecules in solutions, Izvest. Akad. Nauk SSSR, Ser. Fiz. 20(4):397, 1956.

V. L. Levshin, Problems of luminescence of crystallophosphors (conference in Tartu), Vest. Akad. Nauk SSSR (9):99, 1956.

V. L. Levshin, Introductory spectra (to the Fourth Conference on Luminescence, June 1955), Izvest. Akad. Nauk SSSR, Ser. Fiz. 20(4):379, 1956.

V. L. Levshin, Luminescence and Its Technical Uses, Moscow, Izd. AN SSSR, 48 pp., 1956.

V. L. Levshin, V. F. Tunitskaya, and A. A. Cherepnev, The origin of localization levels in ZnS-Cu, Co phosphors, Optika i Spektroskopiya 1(2):255, 1956.

V. L. Levshin and E. G. Baranova, The nature of concentration effects in rhodamine solutions, Izvest. Akad. Nauk SSSR, Ser. Fiz. 20(4):424, 1956.

V. L. Levshin, The P. N. Lebedev Physics Institute of the Academy of Sciences of the USSR, Bol'shaya Sov'etskaya Éntsiklopediya, second edition, Vol. 45, p. 74, 1956.

V. L. Levshin, Influence sur la loi de déclin de la phosphorescence de l'existence de plusieurs systemes de pièges et du phénomena de recapture, J. phys. rad. 17(8-9):684, 1956.

Z. L. Morgenshtern and I. P. Shchukin, Luminescence of color centers in CsI-Tl crystals, Optika i Spektroskopiya 1(2):190, 1956.

A. S. Selivarenko, The excitation state of an imperfect molecular crystal, Izvest. Akad. Nauk SSSR, Ser. Fiz. 20(4):383, 1956.

Z. A. Trapeznikova and V. V. Shchenko, Some optical properties of zinc sulfide phosphors activated with rare earth elements, Doklady Akad. Nauk SSSR 106(2):230, 1956.

L. A. Tumerman, Electroluminescence of organic compounds, Izvest. Akad. Nauk SSSR, Ser. Fiz. 20(5):552, 1956.

V. F. Tunitskaya, Effect of the interaction of activators on the luminescence properties of CaS-Bi, Mn phosphors, in: Trudy FIAN, Vol. 7, Moscow, Izd. AN SSSR, p. 107, 1956.

S. I. Vavilov, "Hot" and "cold" light (thermal radiation and luminescence), Moscow, Znanie, 48 pp., 1956; Bibliography: List of Recommended Books on Problems of Luminescence; also, Collected Works, Vol. IV, p. 113.

L. A. Vinokurov, Action of infrared light on the luminescence of ZnS-Cu, Co and ZnS-Cu, Ni phosphors with varying distribution of electrons over the capture levels, Optika i Spektroskopiya 1(7):901, 1956.

L. A. Vinokurov and M. V. Fok, The quenching of ZnS-Cu, Co and ZnS-Cu, Ni phosphors by infrared light, Optika i Spektroskopiya 1(2):248, 1956.

N. D. Zhevandrov, Polarization spectra of some naphthylamines and polyenes, Izvest. Akad. Nauk SSSR, Ser. Fiz. 20(5):570, 1956.

N. D. Zhevandrov, Polarization diagrams of luminescence of molecular single crystals, Izvest. Akad. Nauk SSSR, Ser. Fiz. 20(5):553, 1956.

1957

M. N. Alentsev, Dependence of the yield of luminescence of ZnS-Cu phosphor on the wavelength of exciting light (report to the Fifth Conference on Luminescence (Crystallophosphors), Tartu, 1956), Izvest. Akad. Nauk SSSR, Ser. Fiz. 21(4):539, 1957.

V. V. Antonov-Romanovskii, Introductory speech to the Fifth Conference on Luminescence (Crystallophosphors), Tartu, 1956, Izvest. Akad. Nauk SSSR, Ser. Fiz. 21(5):715, 1957.

V. V. Antonov-Romanovskii, The diffusion theory of phosphorescence, Optika i Spektroskopiya 3(6):592, 1957.

V. V. Antonov-Romanovskii and M. D. Galanin, On the theoretical derivation of the law of damping of luminescence during resonance quenching, Optika i Spektroskopiya 3(4):389, 1957.

V. V. Antonov-Romanovskii, New results in the study of phosphorescence (report to the Fifth Conference on Luminescence (Crystallophosphors), Tartu, 1956), Izvest. Akad. Nauk SSSR, Ser. Fiz. 21(4):484, 1957.

L. M. Belyaev, M. D. Galanin, Z. L. Morgenshtern, and Z. A. Chizhikova, Dependence of the yield of γ- and photoluminescence of alkali iodides, activated with thallium, on the concentration of activator (report to the Fifth Conference on Luminescence (Crystallophosphors), Tartu, 1956), Izvest. Akad. Nauk SSSR, Ser. Fiz. 21(4):548, 1957.

E. E. Bukke, L. A. Vinokurov, V. E. Oranovskii, Z. A. Trapeznikova, and V. S. Trofimov, Some results of an investigation into electroluminescence (report to the Fifth Conference on Luminescence (Crystallophosphors), Tartu, 1956), Izvest. Akad. Nauk SSSR, Ser. Fiz. 21(5):716, 1957.

E. E. Bukke, Determination of the sign of photoconductivity carriers in phosphors based on ZnS (synopsis of report to the Fifth Conference on Luminescence (Crystallophosphors), Tartu, 1956), Izvest. Akad. Nauk SSSR, Ser. Fiz. 21(5):648, 1957.

E. E. Bukke, Determination of the sign of photocurrent carriers in phosphors based on ZnS, Optika i Spektroskopiya 3(4):334, 1957.

A. A. Cherepnev, The role of physicochemical systems in zinc sulfide luminophors (report to the Fifth Conference on Luminescence (Cryslallophosphors), Tartu, 1956), Izvest. Akad. Nauk SSSR, Ser. Fiz. 21(5):674, (1957).

A. A. Cherepnev, The role of physicochemical systems in zinc sulfide luminophors, in: Reports of the Fifth Conference on Luminescence (Crystallophosphors), Tartu, June 25-30, 1956, p. 52, 1957.

A. A. Cherepnev, Electroluminescing zinc sulfide activated with copper, Optika i Spektroskopiya 2(6):770, 1957.

M. V. Fok, An investigation of the afterglow of the Eu^{+++} ion in a phosphor based on thorium oxide, Optika i Spektroskopiya 2(1):127, 1957.

M. V. Fok, Afterglow of the Eu^{3+} ion in phosphors based on thorium oxide (report to the Fifth Conference on Luminescence (Crystallophosphors), Tartu, 1956), Izvest. Akad. Nauk SSSR, Ser. Fiz. 21(4):505, 1957.

M. V. Fok, The recombination interaction of centers of blue and green luminescence in ZnS-Cu phosphor, Optika i Spektroskopiya 2(4):475, 1957.

M. D. Galanin, International colloquium on luminescence, Priroda (2):106, 1957.

M. D. Galanin, An investigation of scintillators for the recording of nuclear radiations, Vest. Akad. Nauk SSSR (3):130, 1957.

N. A. Garbacheva, Photoluminophors based on phosphates, in: Reports of the Fifth Conference on Luminescence (Crystallophosphors), Tartu, June 25-30, 1956, p. 206, 1957.

N. A. Gorbacheva, Photoluminophors based on phosphates (report to the Fifth Conference on Luminescence (Crystallophosphors), Tartu, 1956), Izvest. Akad. Nauk SSSR, Ser. Fiz. 21(5):682, 1957.

L. T. Kantardzhyan, The relationship between fluorescence and phosphorescence in boron phosphors activated by esculine and uranine as a function of the temperature, Optika i Spektroskopiya 2(3):378, 1957.

M. A. Konstantinova-Shlezinger, Ionic radii of activators and concentration of the latter in crystallophosphors (report to the Fifth Conference on Luminescence (Crystallophosphors), Tartu, 1956), Izvest. Akad. Nauk SSSR, Ser. Fiz. 21(5):673, 1957.

M. A. Konstantinova-Shlezinger, Ionic radii of activators and the concentration of the latter in crystallophosphors, in: Reports of the Fifth Conference on Luminescence (Crystallophosphors), Tartu, June 25-30, 1956, p. 43, 1957.

Yu. S. Leonov, A crystallophosphor lithium-magnesium-tungstate with a manganese activator, in: Reports of the Fifth Conference on Luminescence (Crystallophosphors), Tartu, 1956, Tartu, p. 173, 1957.

Yu. S. Leonov, A crystallophosphor lithium-magnesium-tungstate with manganese activator (report to the Fifth Conference on Luminescence (Crystallophosphors), Tartu, 1956), Izvest. Akad. Nauk SSSR, Ser. Fiz. 21(5):686, 1957.

Yu. S. Leonov, Zinc borate phosphors and their luminescence properties, Doklady Akad. Nauk SSSR 114(5):976, 1957.

A. M. Leontovich, Absorption spectra of plutonium salt crystals, Optika i Spektroskopiya 2(6):696, 1957.

V. L. Levshin and V. F. Tunitskaya, Effect of the wavelength of exciting light and the nature of capture levels of ZnS-Cu, Co phosphors on their filling, Optika i Spektroskopiya 2(3):350, 1957.

V. L. Levshin and V. F. Tunitskaya, Nature of localization levels of ZnS-Cu, Co phosphors and their filling under various conditions of excitation (report to the Fifth Conference on Luminescence (Crystallophosphors), Tartu, 1956), Izvest. Akad. Nauk SSSR, Ser. Fiz. 21(5):695, 1957.

V. L. Levshin, Introductory speech to the Fifth Conference on Luminescence (Crystallophosphors), Tartu, 1956, Izvest. Akad. Nauk SSSR, Ser. Fiz. 21(4):475, 1957.

V. L. Levshin, The contribution of Soviet science to the study of luminescence, Optika i Spektroskopiya 3(5):417, 1957.

V. L. Levshin, N. S. Borodin, and G. P. Neronova, The luminescence of ZnS-Mn phosphors under excitation* (synopsis of a report to the Fifth Conference on Luminescence (Crystallophosphors), Tartu, 1956), Izvest. Akad. Nauk SSSR, Ser. Fiz. 21(4):499, 1957.

V. L. Levshin, N. S. Borodin, and G. P. Neronova, The luminescence of ZnS-Mn phosphors under excitation,* Optika i Spektroskopiya 3(3):258, 1957.

V. L. Levshin and B. D. Ryzhikov, The formation of localization levels of ZnS-Mn phosphors* (report to the Fifth Conference on Luminescence (Crystallophosphors), Tartu, 1956), Izvest. Akad. Nauk SSSR, Ser. Fiz. 21(5):696, 1957.

V. L. Levshin, E. I. Panasyuk, and L. A. Pakhomycheva, Use of the method of labeled atoms to study the volatilization of activators of crystallophosphors during the roasting of a charge, Zhur. Anal. Khim. 12(6):723, 1957.

V. L. Levshin, E. I. Panasyuk, and L. A. Pakhomycheva, Luminescing compositions of constant action with artificially radioactive β-isotopes (report to the Fifth Conference on Luminescence (Crystallophosphors),

Tartu, 1956), Izvest. Akad. Nauk SSSR, Ser. Fiz. 21(4):612, 1957.

V. L. Levshin, The work of S. I. Vavilov in the field of optics, Trudy Inst. Istorii Estestvozn. i Tekhn., Akad. Nauk SSSR, 1957.

G. V. Maksimova, Factors causing the features of crystallophosphors based on cadmium sulfate, in: Reports of the Fifth Conference on Luminescence (Crystallophosphors), Tartu, June 25-30, 1956, Tartu, p. 197, 1957.

A. A. Manenkov, A. M. Prokhorov, Z. A. Trapeznikova, and M. V. Fok, Use of the paramagnetic resonance method to study the state of the activator in phosphors (report to the Fifth Conference on Luminescence (Crystallophosphors), Tartu, 1956), Izvest. Akad. Nauk SSSR, Ser. Fiz. 21(5):779, 1957.

A. A. Manenkov, A. M. Prokhorov, Z. A. Trapeznikova, and M. V. Fok, Use of the paramagnetic resonance method to study the state of the activator in phosphors, Optika i Spektroskopiya 2(4):470, 1957.

Z. L. Morgenshtern and I. P. Shchukin, Luminescence of color centers in CsI-Tl crystals (report to the Fifth Conference on Luminescence (Crystallophosphors), Tartu, 1956), Izvest. Akad. Nauk SSSR, Ser. Fiz. 21(4):593, 1957.

Z. L. Morgenshtern, Spectral distribution of the yield and the absolute yield of photoluminescence of alkali iodides activated with thallium (report to the Fifth Conference on Luminescence (Crystallophosphors), Tartu, 1956), Izvest. Akad. Nauk SSSR, Ser. Fiz. 21(4):544, 1957.

I. K. Plyavin', Kinetics of γ-scintillations in sodium iodide crystals activated with thallium (report to the Fifth Conference on Luminescence (Crystallophosphors), Tartu, 1956), Izvest. Akad. Nauk SSSR, Ser. Fiz. 21(4):549, 1957.

I. K. Plyavin', Kinetics of γ-scintillations in sodium iodide crystals activated with thallium, Optika i Spektroskopiya 2(3):384, 1957.

A. S. Selivanenko, The exciton state of an imperfect molecular crystal, Zhur. Eksptl. i Teoret. Fiz. 32(1):75, 1957.

Z. A. Trapeznikova, The nature of luminescence centers in zinc sulfide luminophors, in: Reports of the Fifth Conference on Luminescence (Crystallophosphors), Tartu, June 25-30, 1956, Tartu, p. 17, 1957.

Z. A. Trapeznikova, The nature of luminescence centers in zinc sulfide luminophors (report to the Fifth Conference on Luminescence (Crystallophosphors), Tartu, 1956), Izvest. Akad. Nauk SSSR, Ser. Fiz. 21(5):678, 1957.

L. A. Tumerman, Duration of the excited state and the quantum yield of fluorescence of chlorophyll in vitro and in vivo, Doklady Akad. Nauk SSSR 117(4):605, 1957.

L. A. Tumerman, A new optical method of mass-spectroscopy (report to the Tenth All-Union Conference on Spectroscopy, July, 1956), Fiz. Sb. L'vovsk. Gos. Univ. (3):81, 1957.

V. N. Varfolomeev and N. D. Zhevandrov, Polarization diagrams of luminescence of stilbene single crystals, Doklady Akad. Nauk SSSR 115(6):1115, 1957.

L. A. Vinokurov and M. V. Fok, Quenching of ZnS-Cu, Co and ZnS-Cu, Ni phosphors by infrared light (report to the Fifth Conference on Luminescence (Crystallophosphors), Tartu, 1956), Izvest. Akad. Nauk SSSR, Ser. Fiz. 21(4):538, 1957.

N. D. Zhevandrov and V. P. Nikolaev, Determination of molecular volumes in solutions by polarization luminescence, Doklady Akad. Nauk SSSR 113(5):1025, 1957.

Yu. N. Zhvanko, Z. L. Morgenshtern, and L. M. Shamovskii, An investigation of the properties of KI-In and KI-Ga phosphors, Optika i Spektroskopiya 2(6):821, 1957.

Yu. N. Zhvanko, Z. L. Morgenshtern, and L. M. Shamovskii, An investigation of the properties of KI-In and KI-Ga phosphors (report to the Fifth Conference on Luminescence (Crystallophosphors), Tartu, 1956), Izvest. Akad. Nauk SSSR, Ser. Fiz. 21(5):752, 1957.

1958

M. N. Alentsev, The relationship between luminescence spectra and absorption of complex molecules, Optika i Spektroskopiya 4(5):490, 1958.

M. N. Alentsev and E. I. Panasyuk, The absorption spectrum of ZnS single crystals, Optika i Spektroskopiya 5(2):207, 1958.

M. N. Alentsev and L. A. Pakhomycheva, Relationship between luminescence and absorption spectra of an aqueous solution of fluorescein (report to the Sixth Conference on Luminescence (Molecular Luminescence Analysis), Leningrad, February 1958), Izvest. Akad. Nauk SSSR, Ser. Fiz. 22(11):1377, 1958.

V. V. Antonov-Romanovskii, On the article by F. I. Vergunas and L. R. Krasovskaya, Optika i Spektroskopiya 5(4):484, 1958.

T. P. Belikova and M. D. Galanin, The mechanism of energy transfer in scintillation plastics (report to the First All-Union Conference on the Synthesis and Study of Scintillation for the Recording of Nuclear Radiation, Moscow, October 1956), Izvest. Akad. Nauk SSSR, Ser. Fiz. 22(1):48, 1958.

E. E. Bukke, L. A. Vinokurov, and M. V. Fok, Effect of the stored light sum on the brightness of electroluminescence of ZnS-Cu, Al phosphor, Optika i Spektroskopiya 5(2):172, 1958.

E. E. Bukke, L. A. Vinokurov, and M. V. Fok, Effect of the stored light sum on the relaxation of brightness of electroluminescence, Inzh.-Fiz. Zhur. 1(7):113, 1958.

L. D. Derkacheva, A study of the effect of the position of substituents on the intensity and frequency of absorption of some naphthalene derivatives, Optika i Spektroskopiya 5(5):542, 1958.

A. A. Dunina, Z. L. Morgenshtern, and L. M. Shamovskii, Absorption and luminescence spectra of alkali halide single crystals activated by indium, Optika i Spektroskopiya 4(1):105, 1958.

M. Fok, Äussere Tilgung, Aktivatorenwechselwirkung und Wanderung der Löcher in den ZnS-Cu and ZnS-Cu, Co phosphoren, in: Halbleiter und Phosphore, Berlin, Braunschweig, p. 593, 1958.

M. V. Fok, External quenching, interaction of activators and migration of holes in ZnS-Cu and ZnS-Cu, Co phosphors, Postepy Fiz. 9(2):197, 1958.

S. A. Fridman and V. V. Shchaenko, Cathodoluminescence of zinc sulfide and zinc cadmium sulfide luminophors activated by rare earth elements, in: Reports of the Seventh Conference on Luminescence (Crystallophosphors), Moscow, 1958, Tartu, p. 288, 1959.

M. D. Galanin and Z. A. Chizhikova, The relationship between the kravets integral and the duration of the excited state of molecules (report to the Eleventh All-Union Conference on Theoretical Spectroscopy, Moscow, December 1957), Izvest. Akad. Nauk SSSR, Ser. Fiz. 22(9):1043, 1958.

M. D. Galanin, Reasons for the dependence of the yield of luminescence of organic substances on the energy of ionizing particles, Optika i Spektroskopiya 4(6):758, 1958.

M. D. Galanin and Z. A. Chizhikova, The quenching of luminescence of organic substances during excitation by α-particles, Optika i Spektroskopiya 4(2):196, 1958.

A. N. Georgobiani and M. V. Fok, An investigation of relaxation processes during electroluminescence, Optika i Spektroskopiya 5(2):167, 1958.

M. A. Konstantinova-Shlezinger, Crystallophosphors with a heterodesmic structure (reports of the Seventh Conference on Luminescence (Crystallophosphors), Moscow, July 1958), Izvest. Akad. Nauk SSSR, Ser. Fiz. 23(11):1304, 1959.

M. A. Konstantinova-Shlezinger, Luminescence analysis and ways for its development, in: Methods of Luminescence Analysis (reports of the Eighth Conference on Luminescence, October 19-24, 1959), Minsk, Izd. AN BSSR, p. 8, 1960.

M. A. Konstantinova-Shlezinger, V. V. Osiko, and L. S. Ulanovskaya, Zinc-lithium-silicate luminophor activated with manganese, Zhur. Neorgan. Khim. 3(6):1286, 1958.

Yu. S. Leonov, A study of the reaction between zinc oxide and boric anhydride using luminescence observations, Zhur. Neorgan. Khim. 3(5):1245, 1958.

V. L. Levshin, A study of luminescence phenomena and the development of its applications in the Soviet Union, Uspekhi Fiz. Nauk 64(1):55, 1958.

V. L. Levshin, New trends in the study of luminescence and the development of its applications, Vestnik Akad. Nauk SSSR (7):26, 1958; Rumanian translation, Analele Rom.-Sov. 13(2):137, 1959.

V. L. Levshin, M. A. Konstantinova-Shlezinger, and Z. A. Trapeznikova, The use of rare-earth elements in the chemistry of luminophors, in: Rare-Earth Elements, Moscow, Izd. AN SSSR, p. 314, 1958.

V. L. Levshin and B. D. Ryzhitov, The formation and action of localization levels of ZnS-Mn phosphors,* Optika i Spektroskopiya 4(3):358, 1958.

V. L. Levshin and E. G. Baranova, Various forms of concentration quenching and the possibility of their separation (report to the Eleventh All-Union Conference on Theoretical Spectroscopy, Moscow, December 1957), Izvest. Akad. Nauk SSSR, Ser. Fiz. 22(9):1038, 1958.

A. N. Nikitina, M. D. Galanin, P. M. Aronovich, T. A. Shchegoleva, and B. M. Mikhailov, An investigation of scintillators containing boroorganic compounds (report to the First All-Union Conference on the Synthesis and Investigation of Scintillators for Recording Nuclear Radiations, Moscow, October 1956), Izvest. Akad. Nauk SSSR, Ser. Fiz. 22(1):12, 1958.

V. E. Oranovskii and Z. A. Trapeznikova, An investigation of spectra of electro- and photoluminescence of phosphors activated by rare-earth elements, Optika i Spektroskopiya 5(3):302, 1958.

V. E. Oranovskii, Electroluminescence, Priroda (11):17, 1958.

V. V. Osiko, Two forms of luminescence centers of manganese in the phase cadmium-lithium-orthosilicate, Doklady Akad. Nauk SSSR 121(3):507, 1958.

I. K. Plyavin', Duration of photoluminescence of alkali halide crystals activated with Tl or In, Optika i Spektroskopiya 4(2):266, 1958.

A. S. Selivanenko, Kinetic parameters of a free exciton for certain types of molecular crystals, Optika i Spektroskopiya 4(1):122, 1958.

A. S. Selivanenko, The quantum mechanical calculation of scattering of a free exciton on a phonon in a molecular crystal, Optika i Spektroskopiya 4(1):92, 1958.

I. P. Shchukin, The relationship of probabilities of captures and recombination of electrons in a KCl-Tl crystal, Optika i Spektroskopiya 5(2):200, 1958.

Z. A. Trapeznikova, Some optical properties of new luminophors activated by rare-earth elements, Postery Fiz. 9(2):211, 1958.

V. S. Trofimov, Dependence of electroluminescence brightness on voltage, Optika i Spektroskopiya 4(1):113, 1958.

V. N. Varfolomeeva and N. D. Zhevandrov, Spatial distribution of polarization of luminescence of stilbene and tolane crystals, Optika i Spektroskopiya 5(5):571, 1958.

L. A. Vinokurov and M. V. Fok, The effect of external quenching on the recombination interaction of centers of blue and green luminescence in ZnS-Cu phosphors, Inzh.-Fiz. Zhur. 1(2):58, 1958.

L. A. Vinokurov and M. V. Fok, Effect of temperature on the recombination interaction of centers of blue and green luminescence in ZnS-Cu phosphor, Optika i Spektroskopiya 4(1):118, 1958.

A. G. Zavrazhin and A. I. Blazhevich, The afterglow of zinc sulfide cathodoluminophors during excitation by an electron beam of low current intensity, in: Reports of the Seventh Conference on Luminescence (Crystallophosphors), Moscow, 1958, Tartu, p. 316, 1959.

N. D. Zhevandrov, A study of the role of localized and free excitons in the luminescence of molecular crystals by polarization methods (report to the Eleventh All-Union Conference on Theoretical Spectroscopy, Moscow, December 1957), Izvest. Akad. Nauk SSSR, Ser. Fiz. 22(11):1332, 1958.

1959

V. V. Antonov-Romanovskii, Initial stages of the quenching of phosphors with several kinds of levels, Optika i Spektroskopiya 7(3):376, 1959.

V. V. Antonov-Romanovskii, Initial stages of the combustion of phosphors with several kinds of levels, Optika i Spektroskopiya 7(4):524, 1959.

V. V. Antonov-Romanovskii, The relaxation of electroluminescence for small deviations from the stationary state, Optika i Spektroskopiya 6(2):229, 1959.

V. V. Antonov-Romanovskii, The superlinear growth in photoconductivity of a phosphor at the initial stages of excitation, Optika i Spektroskopiya 7(6):827, 1959.

V. V. Antonov-Romanvskii, V. G. Dubinin, A. M. Prokhorov, Z. A. Trapeznikova, and M. V. Fok, Detection of the ionization of Eu^{++} in SrS-Eu, Sm phosphor by the method of paramagnetic absorption, Zhur. Eksptl. i Teoret. Fiz. 37(5):1466, 1959.

V. V. Antonov-Romanovskii, Papers on luminescence published in 1955-1958 in collections; Transactions of the Institute of Physics and Astronomy of the Academy of Sciences of the Est. SSR (Nos. 1, 3, 4, 6, and 7), Optika i Spektroskopiya 6(2):269, 1959.

V. V. Antonov-Romanovskii, On the electroluminescence of powdered zinc sulfide layers, Czech. J. Phys. 9:145, 1959.

É. Ya. Arapova, Preparation and investigation of transparent sublimed films of zinc sulfide, in: Reports of the Seventh Conference on Luminescence (Crystallophosphors), Moscow, 1958, Tartu, p. 358, 1959.

T. P. Belikova, Prolonged luminescence of dibenzyl and diphenylamine crystals during photo- and β-excitation, Optika i Spektroskopiya 6(1):117, 1959.

A. Yu. Borisov and L. A. Tumerman, A new type of fluorometer (Reports of the Sixth Conference on Luminescence, Leningrad, February 1958), Izvest. Akad. Nauk SSSR, Ser. Fiz. 23(1):97, 1959.

A. A. Cherepnev, The physicochemical nature of a silver activator in a zinc sulfide luminophor (report to the Seventh Conference on Luminescence (Crystallophosphors), Moscow, July 1958), Izvest. Akad. Nauk SSSR, Ser. Fiz. 23(11):1334, 1959.

Z. A. Chizhikova, Luminescence and Vavilov-Cherenkov radiation in solutions under the action of γ-rays, Optika i Spektroskopiya 7(2):223, 1959.

Z. A. Chizhikova, Energy yield of luminescence during γ-scintillation in a stilbene crystal, Optika i Spektroskopiya 7(2):276, 1959.

L. D. Derkacheva, N. D. Zhevandrov, and Sh. D. Khan-Magometova, A luminescence method for the determination of small amounts of bacteria, Biofizika 7(1):117, 1959.

Hsü Hsü-yung, Determination of the effective capture and recombination cross sections in crystallophosphors (dissertation for the degree of Candidate of Physical and Mathematical Sciences), Trudy FIAN, Moscow-Leningrad, Vol. 11, Izd. AN SSSR, p. 125, 1959.

L. T. Kantardzhyan, Change in the luminescence spectrum of uranyl as a function of the pH of the solution (Reports of the Sixth Conference on Luminescence, Leningrad, February 1958), Izvest. Akad. Nauk SSSR, Ser. Fiz. 23(1):131, 1959.

M. A. Konstantinova-Shlezinger, Crystallophosphors with a heterodesmic structure (report to the Seventh Conference on Luminescence (Crystallophosphors), Moscow, July 1958), Izvest. Akad. Nauk SSSR, Ser. Fiz. 23(11):1304, 1959.

A. V. Lavrov, The formation and inertia properties of some zinc sulfide luminophors (report to the Seventh Conference on Luminescence, Moscow, July, 1958), Izvest. Akad. Nauk SSSR, Ser. Fiz. 23(11):1351, 1959.

V. L. Levshin and E. G. Baranova, An investigation of the nature of concentration quenching of luminescence of dyes in various solvents and the separation of various forms of quenching, Optika i Spektroskopiya 6(1):55, 1959; French translation, J. chim. phys. 55(11):869, 1958.

V. L. Levshin and B. M. Orlov, An investigation of the thermal activation energy of the optical flash of ZnS-Cu, Pb phosphors,* Optika i Spektroskopiya 7(4):530, 1959.

V. L. Levshin, E. G. Baranova, L. D. Derkacheva, and L. V. Levshin, An investigation of association in concentrated solutions of dyes from absorption and luminescence spectra, in: Thermodynamics and the Structure of Solutions, Moscow, Izd. AN SSSR, p. 275, 1959.

V. L. Levshin, V. B. Gutan, and É. N. Karzhavina, The possibility of recombination processes of luminescence in tungstates and uranyl compounds, Optika i Spektroskopiya 6(3):372, 1959.

V. L. Levshin and Yu. A. Klyuev, Formation of luminescence polymers in concentrated solutions of acridine orange and an investigation of their optical properties (Reports of the Sixth Conference on Luminescence, Leningrad, February 1958),* Izvest. Akad. Nauk SSSR, Ser. Fiz. 23(1):15, 1959.

V. L. Levshin, Kh. I. Mamedov, S. R. Sergienko, and S. D. Pustil'nikova, Fluorescence spectra of aromatic hydrocarbons of the diphenyl series and their oxygen- and sulfur-containing analogs,* Izvest. Akad. Nauk SSSR, Otdel Khim. Nauk (9):1571, 1959.

V. L. Levshin and V. N. Rebane, A comparative study of the storage of light sums and the temperature quenching of ZnS-Ag phosphor during excitation by β-rays and light, Optika i Spektroskopiya 7(2):236, 1959.

Z. L. Morgenshtern, Luminescence of nonactivated CsI single crystals, Optika i Spektroskopiya 7(2):231, 1959.

A. N. Nikitina, M. D. Galanin, G. S. Ter-Sarkizyan, and B. M. Mikhailov, Absorption and luminescence spectra of solutions of some substituted polyenes, Optika i Spektroskopiya 6(3):354, 1959.

V. E. Oranovskii and B. A. Khmelinin, An investigation of the electroluminescence of ZnS-Cu single crystals, Optika i Spektroskopiya 7(4):542, 1959.

V. E. Oranovskii, E. I. Panasyuk, and B. T. Fedyushin, An investigation of electroluminescence of ZnS-CuCl single crystals, Inzh. -Fiz. Zhur. 2(1):39, 1959.

V. V. Osiko, The low-temperature luminescence of zinc oxide in the red part of the spectrum, Optika i Spektroskopiya 7(6):770, 1959.

V. V. Osiko, Features of the crystalline phase of zinc silicate and the luminescence properties of zinc silicate luminophors activated with manganese (Reports of the Seventh Conference on Luminescence (Crystallophosphors), Moscow, July 1958), Izvest. Akad. Nauk SSSR, Ser. Fiz. 23(11):1314, 1959.

I. K. Plyavin', The kinetics of photo- and γ-luminescence in some alkali halide crystals activated with Tl, Optika i Spektroskopiya 7(1):71, 1958.

Yu. M. Popov and V. P. Shabanskii, The effect of nonradiative recombination on saturation during cathodoluminescence, Optika i Spektroskopiya 6(6):769, 1959.

Yu. M. Popov, The effect of fast electrons on the stored light sum during cathodoluminescence, Optika i Spektroskopiya 7(5):697, 1959.

Yu. M. Popov, Possible energy losses during cathodoluminescence, in: Reports of the Seventh Conference on Luminescence (Crystallophosphors), Moscow, 1958, Tartu, p. 281, 1959.

Yu. M. Popov, Dependence of the stored light sum at levels of different depth on the density of excitation, Optika i Spektroskopiya 6(6):764, 1959.

I. P. Shchukin, Recombination luminescence and color of KCl-Tl phosphor (dissertation for degree of Candidate of Physical and Mathematical Sciences), Trudy FIAN, Moscow-Leningrad, Vol. 11, Izd. AN SSSR, p. 65, 1959. Appendix: Energy yield of formation of F-centers in KCl-Tl and KCl-Ag during γ-excitation.

Z. A. Trapeznikova, Interaction of "blue" and "samarium" centers in ZnS-Sm (Cl) phosphor, Optika i Spektroskopiya 6(4):512, 1959.

Z. A. Trapeznikova, Luminescence centers in zinc sulfide phosphors activated with samarium (report to the Seventh Conference on Luminescence (Crystallophosphors), Moscow, July 1958), Izvest. Akad. Nauk SSSR, Ser. Fiz. 23(11):1319, 1959.

L. A. Tumerman, The use of spectroscopy in biology and biochemistry, Uspekhi Fiz. Nauk 68(1):93, 1959.

L. A. Tumerman and B. A. Chayanov, Luminescence of dielectrics in alternating electric fields of high intensity, in: Investigations in Experimental and Theoretical Physics, in Memory of Grigorii Samuilovich Landsberg, Moscow, Izd. AN SSSR, p. 231, 1959.

E. G. Vasil'eva and S. A. Fridman, Experiments on the use of thermography to study zinc sulfide (report to the Seventh Conference on Luminescence (Crystallophosphors), Moscow, July 1958), Izvest. Akad. Nauk SSSR, Ser. Fiz. 23(11):1347, 1959.

V. S. Vavilov, B. M. Vul, G. N. Galkin, and S. A. Fridman, The operation of "atomic" sources of current with double energy transformation, Fizika Tverdogo Tela 1(5):826, 1959.

L. A. Vinokurov and M. V. Fok, The simultaneous action of photo- and electroexcitation on electroluminophors, Optika i Spektroskopiya 7(2):241, 1959.

N. D. Zhevandrov, V. I. Gribkov, and V. N. Varfolomeeva, The dependence of polarization of fluorescence of molecular crystals on the wavelength of the radiation (Reports of the Sixth Conference on Luminescence, Leningrad, February 1958), Izvest. Akad. Nauk SSSR, Ser. Fiz. 23(1):57, 1959.

1960

M. N. Alentsev and L. A. Pakhomycheva, The relationship between absorption spectra and luminescence of complex molecules (Report on the Eighth Conference on Luminescence (Molecular Luminescence and Luminescence Analysis), Minsk, October 19-24, 1959), Izvest. Akad. Nauk SSSR, Ser. Fiz. 24(6):734, 1960.

V. V. Antonov-Romanovskii, Stationary luminescence of phosphors in the presence of several kinds of traps, Optika i Spektroskopiya 8(1):73, 1960.

V. V. Antonov-Romanovskii, On the initial stages of luminescent buildup and decay in phosphors with traps of several types, in: Zur Physik und Chemie der Kristallphosphore, Berlin, Akad. Verl., p. 278, 1960.

V. V. Antonov-Romanovskii, Some experimental results on electroluminescent zinc sulfide powders and single crystals, in: Solid State Physics in Electronics and Telecommunications, Proceedings of an international conference held in Brussels, June 2-7, 1958; Vol. 4, Magnetic and Optical Properties, Pt. 2, New York, Academic Press, p. 653, 1960.

É. Ya. Arapova, E. G. Baranova, V. L. Levshin, T. V. Timofeeva, A. K. Trofimov, and P. P. Feofilov, A luminescent method for the quantitative determination of gadolinium in metallic beryllium, Trudy Komissii Anal. Khim. 12:344, 1960.

A. I. Blazhevich, A. G. Zavrazhin, and A. V. Lavrov, The properties of ZnS-Mn, Ni, Cl phosphors during cathodic excitation, Optika i Spektroskopiya 8(4):550, 1960.

A. Yu. Borisov, A high-sensitivity modulation (electron) photometer, Optika i Spektroskopiya 9(1):115, 1960.

L. D. Derkacheva, Change in the fluorescence of naphthalene derivatives as a function of the concentration of hydrogen ions in solution, Optika i Spektroskopiya 9(2):209, 1960.

V. G. Dubinin and Z. A. Trapeznikova, Use of electron paramagnetic resonance to study phosphors based on SrS activated with Eu, Optika i Spektroskopiya 9(4):472, 1960.

V. G. Dubinin, A paramagnetic resonance investigation of the effect of fusion on the state of the activator in a phosphor, Optika i Spektroskopiya 9(4):531, 1960.

V. G. Dubinin and Z. A. Trapeznikova, An electron resonance investigation of the change of valence of an activator during the excitation of SrS-Eu, Sm phosphor, Optika i Spektroskopiya 9(3):360, 1960.

A. A. Dunina, Z. L. Morgenshtern, and L. M. Shamovskii, Absorption and luminescence spectra of alkali halide single crystals activated with indium, in: Scintillators and Scintillation Materials, Moscow, p. 35, 1960.

M. V. Fok, Electroluminescence, Uspekhi Fiz. Nauk 72(2):260, 1960.

Sh. D. Khan-Magometova, N. D. Zhevandrov, and V. I. Gribkov, The effect of β-irradiation on the photoluminescence of molecular crystals (report to the Eighth Conference on Luminescence (Molecular Luminescence and Luminescence Analysis), Minsk, October 19-24, 1959), Izvest. Akad. Nauk SSSR, Ser. Fiz. 24(5):561, 1960.

Yu. A. Kurskii and A. S. Selivanenko, The theory of quenching of luminescence in liquid solutions, Optika i Spektroskopiya 8(5):643, 1960.

Yu. S. Leonov, The technology of luminophors (nonstoichiometric composition of the charge and the fusion during roasting), Zhur. Priklad. Khim. 33(4):769, 1960.

Yu. S. Leonov, A new luminophor, $2Li_2OWO_3(U)$, Optika i Spektroskopiya 9(2):275, 1960.

V. L. Levshin, Introductory speech to the Eighth Conference on Luminescence (Molecular Luminescence and Luminescence Analysis), Minsk, October 19-24, 1959, Izvest. Akad. Nauk SSSR, Ser. Fiz. 24(5):488, 1960.

V. L. Levshin, An investigation of the luminescence of crystallophosphors, Vestnik Akad. Nauk SSSR (9):107, 1960.

V. L. Levshin and N. Kh. Faizi, An investigation of the energy of thermal activation of a flash and localization levels in phosphors based on CaS,* Optika i Spektroskopiya 8(6):875, 1960.

V. L. Levshin, É. Ya. Arapova, and E. G. Baranova, The determination of small amounts of gadolinium, samarium, and europium in metallic thorium, Trudy Komissii Anal. Khim., Akad. Nauk SSSR 12:393, 1960.

V. L. Levshin and V. F. Tunitskaya, Luminescence processes of ZnS-Mn phosphors at the moment of excitation and their kinetics, Optika i Spektroskopiya 9(2):223, 1960.

V. L. Levshin and V. F. Tunitskaya, Thermoluminescence and localization levels of ZnS-Mn phosphors, Optika i Spektroskopiya 8(5):663, 1960.

Z. L. Morgenshtern, The luminescence of nonactivated CsI crystals. II, Optika i Spektroskopiya 8(5):672, 1960.

V. V. Osiko and G. V. Maksimova, The valence of manganese activator in crystallophosphors, Optika i Spektroskopiya 9(4):478, 1960.

V. V. Osiko, Phase composition, luminescence properties and the structure of synthetic zinc silicates containing manganese, Zhur. Neorgan. Khim. 5(2):297, 1960.

Yu. M. Popov and A. S. Selivanenko, Luminescence of a free exciton in a molecular crystal, Optika i Spektroskopiya 9(2):260, 1960.

N. N. Vasil'eva and Z. L. Morgenshtern, Luminescence of nonactivated KI crystals, Optika i Spektroskopiya 9(5):676, 1960.

Yu. V. Voronov, Relationship of intensities of luminescence bands of ZnS-Mn and ZnS-Ag, Mn phosphors as a function of the density of cathodo- and photoexcitation, Optika i Spektroskopiya 9(1):108, 1960.

1961

V. V. Antonov-Romanovskii and V. G. Dubinin, An electron paramagnetic resonance investigation of phosphors based on SrS activated with rare earths (report to the Ninth Conference on Luminescence (Crystallophosphors), Kiev, 1960), Izvest. Akad. Nauk SSSR, Ser. Fiz. 25(4):481, 1961.

V. V. Antonov-Romanovskii, Kinetics of luminescence of phosphors with traps of several kinds (report to the Ninth Conference on Luminescence (Crystallophosphors), Kiev, 1960), Izvest. Akad. Nauk SSSR, Ser. Fiz. 25(3):357, 1961.

V. V. Antonov-Romanovskii, Final stages in the quenching of luminescence of phosphors with levels of several kinds, Optika i Spektroskopiya 10(2):182, 1961.

V. V. Antonov-Romanovskii, Some special cases of the kinetics of phosphorescence, Optika i Spektroskopiya 10(2):214, 1961.

V. V. Antonov-Romanovskii, Luminescence curves of phosphors for comparable times of existence of electrons in traps of various kinds, Optika i Spektroskopiya 10(5):644, 1961.

V. V. Antonov-Romanovskii, The application of the diffusion theory to bimolecular reactions, Fizika Tverdogo Tela 3(6):1896, 1961.

É. Ya. Arapova, An investigation of activator-less ZnS sublimate phosphors (report to the Ninth Conference on Luminescence (Crystallophosphors), Kiev, 1960), Izvest. Akad. Nauk SSSR, Ser. Fiz. 25(3):324, 1961.

E. G. Baranova and B. L. Levshin, The character of bond forces in associated 3B and 6Zh rhodamines in aqueous solutions and the effect of concentration and temperature on the course of migration quenching, Optika i Spektroskopiya 10(3):362, 1961.

T. P. Belikova and M. D. Galanin, The kinetics of luminescence of ZnS-Cu during excitation by α-particles and short light pulses (report to the Ninth Conference on Luminescence (Crystallophosphors), Kiev, 1960), Izvest. Akad. Nauk SSSR, Ser. Fiz. 25(3):364, 1961.

E. E. Bukke, Measurement of power losses in an electroluminescence capacitor (report to the Ninth Conference on Luminescence (Crystallophosphors), Kiev, 1960), Izvest. Akad. Nauk SSSR, Ser. Fiz. 25(4):529, 1961.

Z. A. Chizhikova, The luminescence yield of organic substances (dissertation for the degree of Candidate of Physical and Mathematical Sciences), Trudy FIAN, Moscow-Leningrad, Izd. AN SSSR, p. 178, 1961.

L. Drozd and V. L. Levshin, An investigation of the composition of radiation of nonactivated ZnS-CdS phosphors with change in temperature,* Optika i Spektroskopiya 10(6):773, 1961.

L. Drozd and V. L. Levshin, The arrangement of energy levels of ZnS-CdS phosphors,* Optika i Spektroskopiya 11(5):648, 1961.

V. G. Dubinin, A paramagnetic absorption investigation of the effect of fusion on the concentration of activator in a phosphor, Optika i Spektroskopiya 11(4):518, 1961.

M. V. Fok, The relationship between blue and green luminescence bands of ZnS-Cu phosphors during electrical excitation, Optika i Spektroskopiya 11(1):98, 1961.

M. V. Fok, Electroluminescence (brief report of session of a seminar in the P. N. Lebedev Physics Institute of the Academy of Sciences of the USSR in honor of the memory of S. I. Vavilov, March 1961), Uspekhi Fiz. Nauk 75(2):259, 1961.

M. D. Galanin and Z. A. Chizhikova, Duration of photo- and radioluminescence of anthracene-naphthalene crystals, Optika i Spektroskopiya 11(2):271, 1961.

A. N. Georgobiani and M. V. Fok, Dependence of the phase of electroluminescence brightness waves on the parameters of the exciting voltage, Optika i Spektroskopiya 11(1):93, 1961.

A. N. Georgobiani, The excitation of electroluminescence of zinc sulfide, Optika i Spektroskopiya 11(3):426, 1961.

A. N. Georgobiani and M. V. Fok, The process determining the dependence of mean brightness of electroluminescence on the voltage, Optika i Spektroskopiya 10(2):188, 1961.

N. A. Gorbacheva and V. V. Osiko, The valence of Sn and Mn activators in crystallophosphors (report to the Ninth Conference on Luminescence (Crystallophosphors), Kiev, 1960), Izvest. Akad. Nauk SSSR, Ser. Fiz. 25(4):454, 1961.

N. A. Gorbacheva, B. M. Gugel', M. A. Konstantinova-Shlezinger, E. S. Lapir, and T. G. Rutshtein, Phosphate luminophors for luminescent lamps with improved color transmission (report to the Ninth Conference on Luminescence (Crystallophosphors), Kiev, 1960), Izvest. Akad. Nauk SSSR, Ser. Fiz. 25(4):455, 1961.

V. I. Gribkov, N. D. Zhevandrov, and Sh. D. Khan-Magometova, Polarization of luminescence of molecular crystals during excitation by β-rays, Optika i Spektroskopiya 10(4):549, 1961.

N. N. Grigor'ev and Yu. A. Kulyupin, Some results of the investigation of damage in luminophors during luminescence (report to the Ninth Conference on Luminescence (Crystallophosphors), Kiev, 1960), Izvest. Akad. Nauk SSSR, Ser. Fiz. 25(4):526, 1961.

N. N. Grigor'ev and Yu. A. Kulyupin, Some results of the investigation of spoiling of luminophors during electroluminescence, Optika i Spektroskopiya 10(6):780, 1961.

M. A. Konstantinova-Shlezinger, Different types of crystallochemical systems of crystallophosphors and their luminescence properties (report to the Ninth Conference on Luminescence (Crystallophosphors), Kiev, 1960), Izvest. Akad. Nauk SSSR, Ser. Fiz. 25(4):442, 1961.

S. I. Kunenkov, The effect of high pressures on the formation of crystallophosphors and their properties (report to the Ninth Conference on Luminescence (Crystallophosphors), Kiev, 1960), Izvest. Akad. Nauk SSSR, Ser. Fiz. 25(3):419, 1961.

A. V. Lavrov, ZnS-Ag luminophor with boron coactivator, Optika i Spektroskopiya 11(1):128, 1961.

Yu. S. Leonov, Lithium-magnesium-tungstate activated with uranium, Optika i Spektroskopiya 10(5):679, 1961.

V. L. Levshin and B. D. Ryzhikov, The effect of dimensions of natural and pulverized crystals on the luminescence of zinc sulfide phosphors* (report to the Ninth Conference on Luminescence (Crystallophosphors), Kiev, 1960), Izvest. Akad. Nauk SSSR, Ser. Fiz. 25(3):362, 1961.

V. L. Levshin, Introductory speech to the Ninth Conference on Luminescence (Crystallophosphors), Kiev, 1960, Izvest. Akad. Nauk SSSR, Ser. Fiz. 25(1):2, 1961.

V. L. Levshin and B. D. Ryzhikov, Dependence of the yield and other optical properties of zinc sulfide phosphors on the dimensions of crystals which have not been pulverized,* Optika i Spektroskopiya 10(4):505, 1961.

V. L. Levshin, Notes on the concepts "yield," "mean duration," and "law of quenching" and their application, Optika i Spektroskopiya 11(3):362, 1961.

V. L. Levshin and P. A. Pipinis, A study of the capture levels of CaS phosphors by methods of exoelectron emission and thermoluminescence* (report to the Ninth Conference on Luminescence (Crystallophosphors), Kiev, 1960), Izvest. Akad. Nauk SSSR, Ser. Fiz. 25(4):471, 1961.

V. L. Levshin, A brief account of scientific, teaching and social activity, in: Sergei Ivanovich Vavilov (1891-1951), second edition, revised, Moscow, Izd. AN SSSR, p. 7, 1961.

V. L. Levshin and B. M. Orlov, A study of the energy of thermal activation of optical quenching of some crystallophosphors* (report to the Ninth Conference on Luminescence (Crystallophosphors), Kiev, 1960), Izvest. Akad. Nauk SSSR, Ser. Fiz. 25(4):466, 1961.

V. L. Levshin, Yu. V. Voronov, V. B. Gutan, S. A. Fridman, and V. V. Shchaenko, A study of the action of double activation by silver and samarium on localization levels and the radiation of zinc sulfide phosphors (report to the Ninth Conference on Luminescence (Crystallophosphors), Kiev, 1960), Izvest. Akad. Nauk SSSR, Ser. Fiz. 25(3):392, 1961.

V. L. Levshin, Development of the idea of S. I. Vavilov in the field of luminescence, Uspekhi Fiz. Nauk 75(2):241, 1961.

V. L. Levshin, A. N. Terenin, and I. M. Frank, Development of S. I. Vavilov's work in physics, Uspekhi Fiz. Nauk 75(2):215, 1961.

V. L. Levshin, Sergei Ivanovich Vavilov (30th anniversary of his birth), Uspekhi Fiz. Nauk 73(3):373, 1961.

V. L. Levshin, Sergei Ivanovich Vavilov (1891-1951), in: Men of Russian Science, Moscow, Fizmatgiz, p. 380, 1961.

V. E. Oranovskii, An investigation of the electroluminescence of single crystals based on ZnS (report to the Ninth Conference on Luminescence (Crystallophosphors), Kiev, 1960), Izvest. Akad. Nauk SSSR, Ser. Fiz. 25(4):516, 1961.

V. V. Osiko, Intercrystalline reactions in crystallophosphors and the structure of luminescence centers, Optika i Spektroskopiya 11(5):642, 1961.

Yu. M. Popov, Features of cathodoluminescence at high densities of excitation (report to the Ninth Conference on Luminescence (Crystallophosphors), Kiev, 1960), Izvest. Akad. Nauk SSSR, Ser. Fiz. 25(3):405, 1961.

A. S. Selivanenko, Times of relaxation of an exciton in molecular crystals, Optika i Spektroskopiya 11(5):694, 1961.

A. S. Selivanenko, Scattering of free excitons by lattice defects of a molecular crystal, Fizika Tverdogo Tela 3(4):1009, 1961.

N. N. Vasil'eva and Z. L. Morgenshtern, Luminescence of nonactivated alkali iodides (report to the Ninth Conference on Luminescence (Crystallophosphors), Kiev, June 20-25, 1960), Izvest. Akad. Nauk SSSR, Ser. Fiz. 25(1):47, 1961.

L. A. Vinokurov and M. V. Fok, Determination of the depth of electron traps in phosphors based on ZnS from the flash under the action of infrared light, Optika i Spektroskopiya 10(3):374, 1961.

L. A. Vinokurov and M. V. Fok, The role of the luminescing action of exciting light in the kinetics of luminescence of ZnS-Cu crystallophosphor, Optika i Spektroskopiya 10(2):225, 1961.

N. D. Zhevandrov, Achievements of the Soviet school of luminescence (Tenth Conference on Luminescence, Moscow, June 26-July 1, 1961), Vestnik Akad. Nauk SSSR (10):130, 1961.

N. D. Zhevandrov, V. I. Gribkov, and Sh. D. Khan-Magometova, The effect of birefringence of exciting light on the polarization of luminescence of molecular crystals, Optika i Spektroskopiya 11(5):629, 1961.

N. V. Zhukova, G. K. Evdokimova, and V. L. Levshin, Dependence of the quenching of phosphorescence on the filling of localization levels of electrons and temperature* (report to the Ninth Conference on Luminescence (Crystallophosphors), Kiev, 1960), Izvest. Akad. Nauk SSSR, Ser. Fiz. 25(4):476, 1961.